TRAVELLER IN A VANISHED LANDSCAPE

David Douglas F.L.S. 1798–1834
enlarged from a pencil drawing ætat 30 by his niece Miss Atkinson

TRAVELLER IN A VANISHED LANDSCAPE

THE LIFE AND TIMES OF DAVID DOUGLAS

BY WILLIAM MORWOOD

GENTRY BOOKS · LONDON

First published in Great Britain 1973
ISBN 0 85614 022 8

Published by Gentry Books Limited
55-61 Moorgate, London, EC2R 6BR
Printed and bound in the United
States of America
Designed by Shari de Miskey

To Minerva with love

CONTENTS

For riches vanish, the most stately mansions fall into decay, the most prolific families die out sooner or later: the mightiest states and the most flourishing kingdoms may be overthrown: but the whole of nature must be obliterated before the genera of plants disappear and he be forgotten who held the torch aloft in botany.

CARL LINNAEUS (1707–1778)

TRAVELLER IN A VANISHED LANDSCAPE

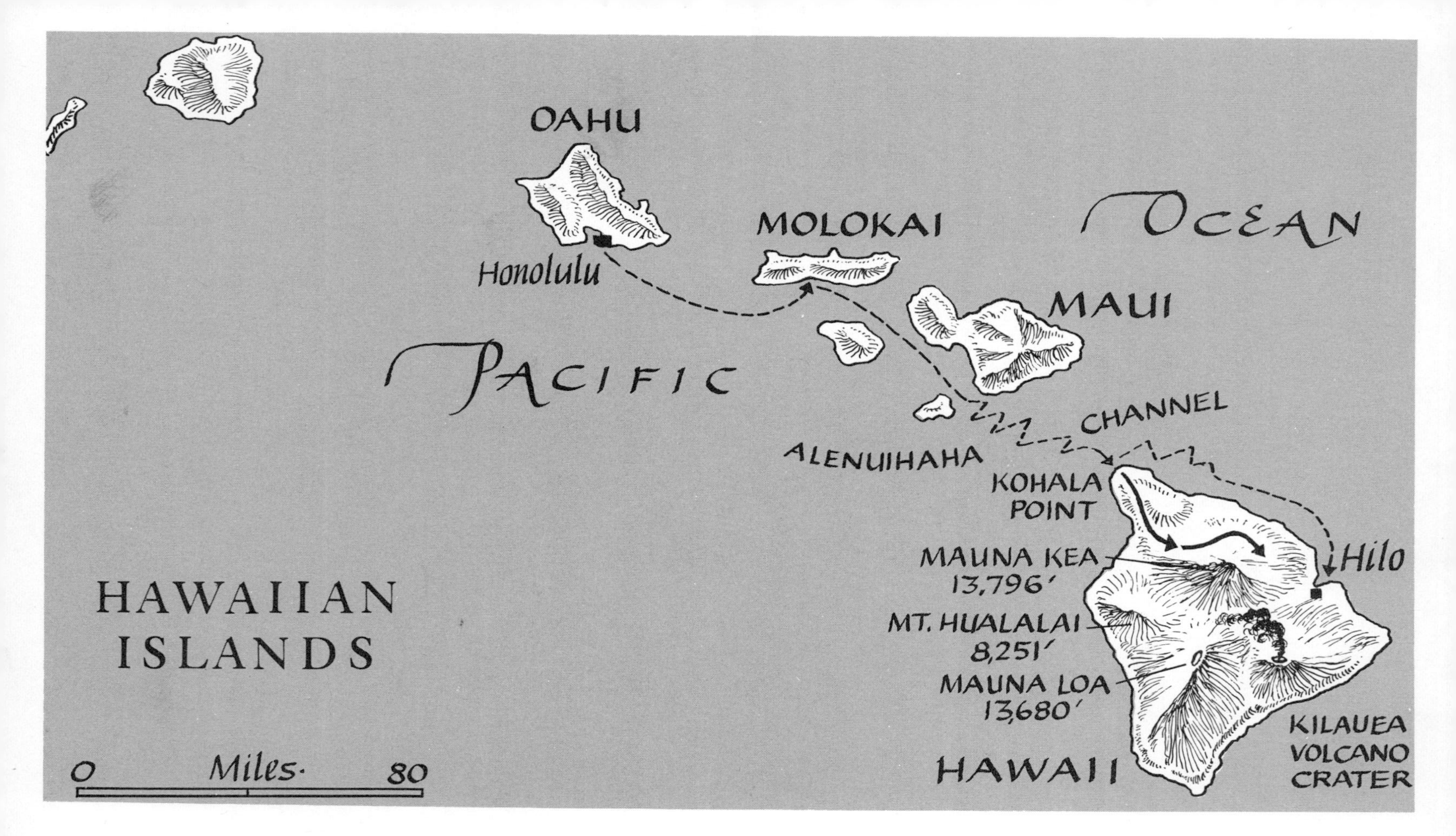
OAHU
Honolulu
MOLOKAI
OCEAN
MAUI
PACIFIC
CHANNEL
ALENUIHAHA
KOHALA
POINT
MAUNA KEA
13,796′
Hilo
MT. HUALALAI
8,251′
MAUNA LOA
13,680′
KILAUEA
VOLCANO
CRATER
HAWAII
HAWAIIAN
ISLANDS
0 Miles 80

PROLOGUE

On July 9, 1834, a coastal schooner plying between ports of the Hawaiian Islands found itself unable to make headway against the boisterous winds and heavy seas of the Alenuihaha Channel, so put in at Kohala Point to take on ballast. This was at the northern tip of Hawaii, the big island, and all on board knew they were in for an uncomfortable three days of beating into the wind before reaching the east coast port of Hilo.

One of the passengers asked to be put ashore. His name was David Douglas, "a sturdy little Scot, handsome rather, with head and face of fine Grecian mould." He had become famous during the past ten years through his amazing botanical expeditions into California and the Columbia River region of the Pacific Northwest. It was to be said of him that "to no single individual is modern horticulture more indebted than to David Douglas" and, in another tribute, "throughout Europe and in the United States of America there is scarcely a spot of ground deserving the name of a Garden which does not owe many of its most powerful attractions to the living roots and seeds . . . sent by him to the Horticultural Society of London."

Douglas had by now been in the Hawaiian Islands (then called the Sandwich Islands) for little more than six months, but already, with almost feverish energy, he had collected more than two thousand botanical specimens and climbed every notable peak of the archipelago.

He asked to be put ashore for two reasons: the prospect of a jarring, monotonous passage to windward was unappealing, and he had heard of an overland trail to Hilo which he wanted to try. It was a route that skirted the extinct volcano of Mauna Kea at a height of six thousand feet, and so promised cooler conditions than the sun-baked deck of a ship. Like many fair-complexioned people, Douglas had an aversion to intense sunlight that bordered on a phobia.

The distance of the overland route, something under a hundred miles, hardly challenged him. In an era of prodigious walkers, Douglas's exploits were notable. He thought nothing of covering fifty miles a day under favorable conditions. Rough terrain, of course, could slow him up, as could his pack which, when fully loaded with instruments for botany, geology, and astronomy, weighed sixty pounds.

On the present trip he intended to travel light. Allowing for an unfamiliar path, with the inevitable detours and misdirections, he still expected to be dockside at Hilo in plenty of time to collect his baggage when the ship came in.

At least that was his expectation before the schooner's captain insisted that John leave the ship with him. John was a black man, the servant of an American missionary. The latter had disembarked at Molokai, an island already famous for its beauty and still unmarred by the presence of a leper colony. There'd been a mixup, and somehow John had remained aboard the schooner. Now the captain seized this opportunity to get rid of him. Douglas may have protested the unwelcome responsibility, but since he was probably traveling as a guest of the Reverend Diell, John's master, and was to be the guest of other American missionaries when he reached Hilo, he had little option but to take the black man along. Douglas's Scotch terrier, Billy, "the companion of all my journeys," completed the landing party.

The two men and the dog were put ashore sometime during the morning of July 9, and they began to climb a winding path to the plateau known as the Kohala Highlands. Possibly they reached level ground that afternoon. If so, John's endurance must have been pushed to the limit. The trail was a crude affair, scuffed and cut from volcanic rock, hard on shoes and feet. Even Douglas, when he'd first encountered the jagged surfaces of Hawaii, had pulled up lame. John must have been limping badly before much time had passed. But limping or not, they couldn't stop for the night until they found water. Though the highlands were annually drenched by more than a hundred inches of rain, precipitation either disappeared through the porous rocks or flowed down the slopes to irrigate the luxurious vegetation along the coast. The plateau was left with barely enough moisture to support sparse grass and ragged brush.

Sooner or later water was found. It may have come from a chance pond tucked in a fold of the terrain; more likely it was rainwater caught in an oilcloth during a shower. Canteens were filled, thirsts slaked, and the residue of liquid, foaming with Douglas's ever-ready Epsom salts, may have served as a medicinal dip for John's ailing feet. Supper was dried meat and fruits brought from the ship. The night was passed under the oilcloth, utilized now as a tent.

Douglas liked to begin a day's journey before dawn, but it is doubtful if he got an early start on the morning of July 10. The condition of John's feet must have called for considerable padding and binding to fit them for travel. The advance, when it got under way, was inevitably slow. We can see the little procession: Billy skipping ahead, sniffing busily; Douglas next, slapping at brush with his cane, irritated by the enforced snail's pace; John, last in line, hobbling, limping, stumbling, desperate to keep up. Long before noon Douglas must have recognized that the situation was hopeless and begun to look for a place where he could decently deposit the black man.

We know little about John except that he had lived in confusion for many months. There had been the long and terrifying sea voyage from New London, Connecticut, followed by the shock of the Sandwich Islands where the people, language, food, and dress must have seemed disordered figments of a nightmare. And now his master, his last link with reality, had somehow disappeared.

We don't know how the unfortunate man came to be employed by the Reverend John Diell, chaplain of the American Seamen's Friend Society, recently arrived in Honolulu. We know little about the chaplain himself and have no clue as to why a spare-living New England missionary undertook to transport a servant of whatever color some sixteen thousand nautical miles to a remote outpost of western civilization. It has been suggested that John was still a slave and it was Diell's intention to free him in a public ceremony, thus bearing witness to the mercy of Providence and, incidentally, to the benevolence of missionaries. But we don't know. John remains a mystery and his connection with David Douglas serves as another incongruous link in what was to prove a bizarre chain of events.

Sometime during the day a hunter's lodge appeared—a circular hut thatched with pili grass, resembling a haystack in texture and contour. Its owner was, predictably, an ex-sailor, British or American, who had jumped ship and climbed the mountain to eke out a living as a cattle hunter. Thousands of the wild animals—said to have sprung from an original pair left by the English explorer Captain George Vancouver in 1793—populated the highlands, though they were rarely to be seen by day. Remorseless hunting had driven them into the cover of brush-filled gullies, from which they emerged at night in search of water and forage.

A cluster of adults and children would have watched Douglas and John approach the lodge. White hunters generally married into large Hawaiian families, and all members willing to work were welcomed to the mountain. Beaters were needed to drive cattle out to be shot, and, later, strong backs must carry hides, tallow, and salted meat down to the coast for sale to ships—usually whalers standing offshore to take on supplies and saleable commodities.

No doubt Douglas made two arrangements with the lodge owner: the first, in John's hearing, for a night's accommodation; the second, later and secretly, for John's board and lodging until he was able to travel again. Money changed hands, a down payment plus an additional sum promised when John was safely delivered to the missionaries in Hilo.

Unquestionably, Douglas decamped early on the morning of July 11, before dawn and before John awoke. Though we can understand the reasons for his stealthy departure—the black man's pleadings and lamentations would have been harrowing—a tragedy might have been averted if Douglas had been more open with his childlike companion, if he had at least tried to convince him that their separation was only for a few days. As it was, we can only guess at John's panic when he found himself abandoned. Did he rush after Douglas till pain and exhaustion overcame him, then crawl into a gulley to die? We don't know. All that's certain is that John was never seen or heard from again.

In the meantime, freed from encumbrances, Douglas pressed his stride, making up for lost time. He had left the ship traveling light, but it became a cause for conjecture and suspicion later when it was discovered just *how* light. No specimen papers, magnifying glass, canisters for seeds—none of the paraphernalia of his profession that was second nature for him to carry was found in his possession. It was as if, on this, his last journey, the man who had shown such "perseverence in almost every species of danger, privation, and hardship for the advancement of knowledge" had lost enthusiasm for the business of his life.

Striding across the highlands, Douglas was encompassed by the vault of an enormous sky that, tucking in around and under him, gave the illusion of progressing through a gigantic blue balloon. Hawaii's three great mountains loomed ahead: Mauna Kea above him, almost on top of him, its vast bulk swelling up from the path on which he walked; Mauna Loa far off to the south, its volcanic cone scarred and jagged; Hualalai, closer at hand, front and right, gray, barren, deeply cleft, completing the unreal, moonlike topography.

Earlier in the year Douglas had climbed Mauna Kea and had written of his sensations upon achieving its summit: "Man feels himself as nothing—as of standing on the verge of another world . . . in the presence of a great and good, and wise and holy God." No doubt he glanced up at the peak occasionally as he moved across its northern shoulder, but it is unlikely that his former sentiments recurred to him. Too many ghosts and fantasies had crowded into his mind to permit metaphysical speculations.

It had been apparent for some time that David Douglas was not at peace with himself. He had alienated old friends by boasts and insults, and alarmed new ones by a garrulity he seemed unable to control. He, who had once been celebrated along the Columbia River for his tact and patience with Indians, had taken to abusing Hawaiian guides and baggage carriers to the extent that none would work for him twice; and a youthful quality, described by an admirer as "singular abstemiousness," had, by the time Douglas became thirty-five, been eroded by steady and often excessive drinking.

The path marched on through the knee-deep brush, a red thread in the gray waste, without trees or boulders to give dimensions or mark a traveler's rate of progress. In that vast and empty space Douglas might as well have been walking on a treadmill or through the foreshortened landscape of a dream. It wasn't until the roadway crooked to the east that the contour of the land changed abruptly. The hillslope lost its languor and collapsed to his left, plunging with often precipitous speed to the ocean. The vegetation around Douglas's feet was still sparse and dry, but below him, in the gulches (indented valleys shaped like creases in a half-opened umbrella), lush greenery reached up. He could catch glimpses far below of coconut groves and taro fields—crops introduced from the Marquesas Islands a thousand years before though now masquerading as natives. Out at sea whitecaps sparkled and perhaps a ship ran with taut canvas before the wind.

A ship would attract Douglas's attention because he knew that he must soon find one bound for England—for England because he had no other place left to go. Back home, and then what? What do you do when, spider-like, you've spun out the entire substance of your life?

How long had it been since he'd written that fateful letter? Long enough, even counting the eight months it took a ship to double the Horn and cross the seas. Time enough for word to emerge from Whitehall and filter out to the most remote of his Majesty's ships. Perhaps it was already too late. His credentials from the Colonial Office might even now be obsolete, invalid for passage anywhere.

Sometime that afternoon on the slopes of Mauna Kea Douglas became aware that the day was ending. It could have been shadows as the sun dipped behind the mountain, or it could have been Billy, still gamely up ahead but looking back with reproachful frequency, that made him conscious of the hour. He cleared his mind of invading thoughts and began looking for accommodations for the night.

He found them at the lodge of a hunter called Davis. No first name was ever recorded for the man, which suggests a dour, suspicious renegade from white society, an impression that was to be borne out later during the investigation. For when asked to speak freely about the night Douglas spent at his lodge, Davis could remember nothing until he learned that another man was under suspicion. Then he invented a cock-and-bull story about seeing a purse stuffed with money, obviously hoping to further incriminate that other man, a complete stranger.

In actual fact, Davis might well have had forebodings when Douglas stopped at his hut on the evening of July 11. What was this well-turned-out Scotsman doing on the mountain slope? Respectable foreigners lived in the coastal towns; there had to be a reason for one to wander. Perhaps ship captains had sent him to locate runaway seamen. Whatever it was, it boded no good for Davis, and the sight of coins in Douglas's hand would have made him no less suspicious.

There was beef for supper, fresh or salted, depending how recently a kill had been made. Inevitably there was poi—then as now the starchy staple of the islanders. Bananas or sections of sugar cane would top off the meal, and Douglas might additionally brew himself tea before retiring. Rainy or clear, he would sleep under his oilcloth outside the hut. Fleas in a hunter's lodge were more than a theoretical possibility.

The hours begin to close in now. If Douglas maintained his schedule of early departures, he would have covered a half-dozen miles after dawn, July 12, before encountering the last known marker of his journey.

Shortly before six o'clock that morning he arrived at a junction of paths above a gulch sonorously called Laupahoehoe. Close by was a lodge, larger and better constructed than any Douglas had yet seen. It was rectangular in shape and boasted a practical door, as well as windows paned with parchment, a change from the usual ventilation holes cut into straw. Despite the early hour, the proprietor was at work outside his establishment, cleaning hides for shipment. His name was Edward Gurney.

We must turn to subsequent events in order to get an impression of Gurney. Unlike Davis and other marginal hunters of the mountain slope, he was a man of organizational ability who ran an efficient and prosperous cattle business. His reputation seems to have penetrated even to the coastal towns, for the Reverend Joseph Goodrich, with whom Douglas was to stay in Hilo, had heard of him. Before meeting Gurney, the Reverend Mr. Goodrich described him somewhat disparagingly as "an Englishman, a convict from Botany Bay, who left a vessel at these islands some years ago." But later the missionary's attitude changed completely and he conveys to us the sense of a strong and resourceful person whose actions in a crisis "greatly relieved" Mr. Goodrich's mind.

On the morning of July 12, puzzled by branching trails, Douglas called on Gurney for directions. No doubt Billy exchanged a fierce fusillade of barks with the hunter's dogs, and for a while both men worked at restoring peace. Finally Douglas asked his question and Gurney pointed out the path, though warning of other forks which branched from the Hilo trail in the miles ahead.

At some point, either Douglas requested or Gurney offered his services as a guide until the route became less ambiguous. Douglas still had a couple of hours of travel ahead of him before his usual breakfast halt, but something happened to change his plans.

It may have been the sight of Irish potatoes. Gurney grew a patch of them behind his house. He'd learned that New England whaling crews yearned for grog, women, and boiled potatoes in that order when they reached port, and he had been able to dispose of all the tubers he could grow, and for fancy prices. It is just possible that Douglas saw a plate of spuds ready for the frying pan and found them irresistible.

Gurney must have been pleased to have Douglas stay for breakfast. There would be the pleasure of speaking English, for one thing—rare for him these days except for his forays to the ships. They may have discussed agricultural crops of the region, then progressed to cattle hunting. Douglas would be interested to learn that Gurney had developed a new technique for capturing his quarry. Where other hunters laboriously beat the bushes to get a shot at the animals, Gurney dug pits, set traps, and waited. He may have scratched out a diagram with charcoal to demonstrate his method.

Water was the bait, either in natural pools or in ponds hollowed out of lava and plastered with mud to retain rain water. A fence was constructed around this water hole, leaving openings through which the animals must pass in order to drink. In each opening a pit, measuring eight feet long by four feet wide by six feet deep, was dug and camouflaged. Once the animal tumbled in, the rocky walls of the pit held it securely.

Breakfast over, Douglas would be impatient to be on his way again. He was still some forty miles from Hilo and would have to step briskly in order to reach the Goodrich home before nightfall.

The men started out. As they walked, Gurney made a suggestion. One of his cattle traps was located just off the path a few miles ahead. If Douglas was interested, he'd be glad to show him the pits. A bull and a cow had been captured during the night while a third trap was still unsprung. Douglas could have a close look at the captives and also examine the artful camouflage that concealed the remaining pit.

Douglas must have readily agreed to the suggestion. The prospect of seeing animals that weighed half a ton straining in captivity would have excited his imagination. Sheer bulk always fascinated him. During the years of his greatest botanical success, he discovered many beautiful and delicate

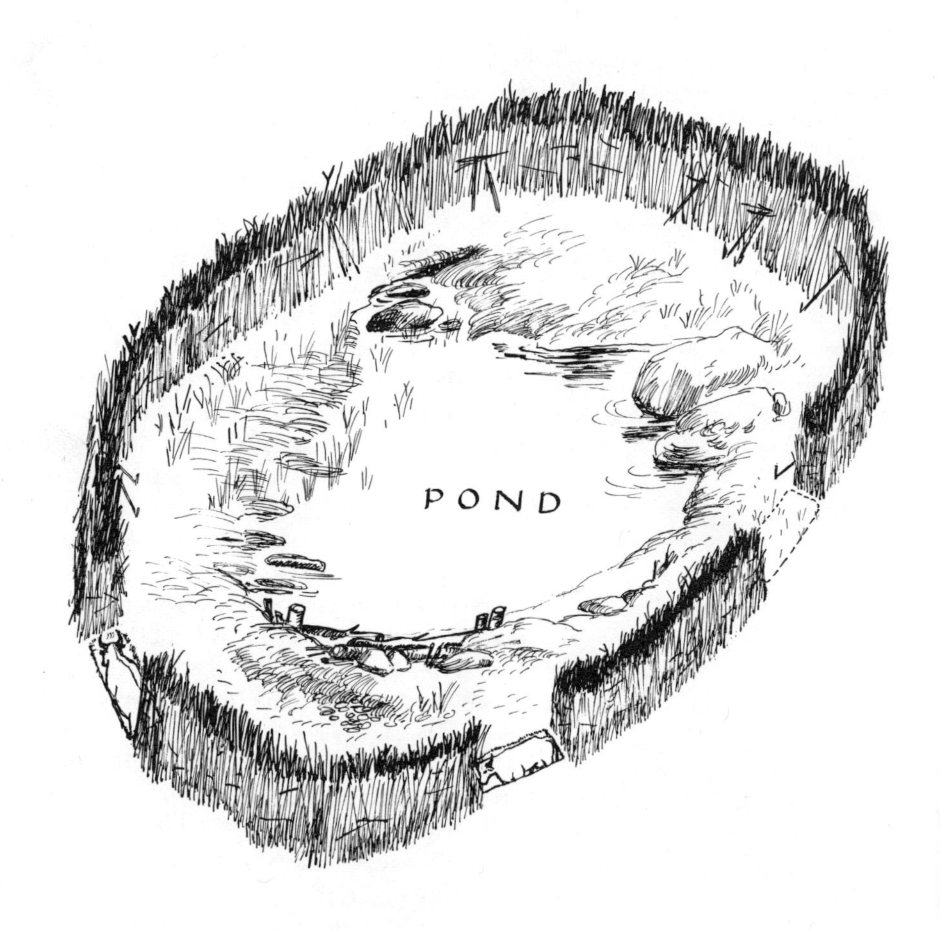

Camouflaged Pits

garden flowers; but the introductions which loomed largest in his mind, literally and figuratively, were the giant trees—the sugar pine, the Sitka spruce, and the towering conifer that bears his name, the Douglas fir.

The men traveled easily for more than a mile, Douglas setting the pace but Gurney matching him without difficulty. The path was still high on the shoulder of Mauna Kea, directly on the eastern slope now, but the flanks of the mountain had flared out, easing the sense of precipitous fall down to the ocean. The scuffed red lava trail stretched before the men like a dirty ribbon, tucking in and out of folds in the hillside. Gurney could point to rock formations miles ahead, and somewhere there was a landmark beneath which the cattle pits were located.

Douglas had been affable all morning, talking, laughing, possibly over-expressive. But suddenly, about two and a half miles from their destination—the calculation is Gurney's—his mood changed. He grew silent and morose, as though weary of companionship. Finally he stopped and held out his hand. He thanked Gurney for having come this far but, as the path ahead seemed uncomplicated now, he preferred to go on alone.

Gurney later described the sudden good-bye as the equivalent of being "dismissed." He was surprised, but doesn't seem to have been resentful of Douglas's tone and manner. Perhaps he'd guessed from various indications that his companion was under some kind of strain that might best be relieved by solitude. But before turning back, he cautioned Douglas again about the cattle traps, if he still meant to visit them. Gurney wasn't concerned about the occupied pits. The snorting and heaving of the captured animals would give ample warning of danger there. He was troubled about the unsprung trap. The camouflage was so convincing, even to hoof marks printed into the mud covering, that unless the area was approached with the greatest caution, a six-foot fall with broken bones could result.

It is doubtful if Douglas acknowledged this well-meant warning, though he may have greeted it with a hoot of laughter. At this late date, after surviving a thousand risks, he wasn't about to break a leg in a primitive Hawaiian cattle pit.

So the men parted. We're left with a picture of Douglas walking rapidly down a winding path, bundle over shoulder, stick in hand, Billy loping ahead. Edward Gurney watches him go and is a trifle uneasy, he doesn't know why. If ever a man was capable of taking care of himself, it's that sturdy Scot.

Gurney returned to the lodge and to the occupation interrupted by Douglas—cleaning hides. He may have speculated for a while about his recent companion with the mercurial temperament, but he gave no thought to the pits. He didn't intend to visit them again until the next morning when, with luck, the third trap would be sprung. For efficiency's sake, Gurney liked to wait for a full house. Then he could move in with a crew to dispatch the animals, flay and dress the carcasses, and reset the traps—a day's work.

He looked up at about eleven o'clock when two of his men came running from the direction of the pits. He could tell from a distance that something was wrong. They came up and stared at him, unable to speak. Then the words came while the eyes, huge and dilated, kept staring.

It was the Kauka, the white doctor, they told him. He must come at once.

1

HORTICULTURE FOR THE SELECT

"DEAR SIR, I HAVE BEEN TURNING MY attention to the formation of a Horticultural Society," began a letter of June 29, 1801.

There could have been no better time to interest the English landed gentry in the scientific cultivation of their gardens. Agriculture was booming as never before or since. The nation, still in a panic over the Malthusian prediction that unchecked population would outrun food supply, was further agitated by Napoleon's threat to starve it into submission with a blockade. As a result, the plowshare rose to almost equal status with the sword—a phenomenon unprecedented in English history. Poetically, at least. Socially, the British have always preferred a man who could run you through the belly to one who could fill it.

The government, buying food to support its allies and storing up supplies for its own coming struggle with Napoleon, helped to push up agricultural prices. Marginal lands were brought under cultivation and intensive efforts made to increase production on fertile acres. Books, pamphlets, periodicals, and itinerant lecturers provided a ceaseless flow of information on the latest agricultural advances. Cash was awarded yeomen for bumper crops of oats or barley; dukes proudly showed off medals for outsized turnips.

Not only were the crops improved and multiplied, but cattle, pigs, sheep, chickens, and turkeys were bred into different shapes and sizes to yield more steaks, bacon, fleece, and larger eggs and drumsticks. In this atmosphere of heady scientific development, the landed gentry were more than ready to experiment with fruits, flowers, and vegetables in their own kitchen and pleasure gardens.

"We wish to clash with no society at present instituted," the letter of 1801 continued. "By this means we shall give no offense to any party."

It was a realistic consideration. The field was already so crowded with organizations devoted to agricultural improvement that it seemed hardly possible to launch a new one without landing it on somebody's toes. True, a distinction was made in a subsequent letter emphasizing that the proposed new society would be "for the sole purpose of encouraging Horticulture"; but the difference between cultivating gardens (horticulture) or fields (agriculture) must have seemed negligible in those unspecialized days.

Still, as anyone who has labored as a founding father knows, the question most often asked about a new organization is not "What's it for?" but "Who's in it?"; and in this respect the promoters of the Horticultural Society of London made few mistakes.

The writer of the two letters was John Wedgwood, a member of the prestigious pottery family of Staffordshire. He enlisted as a charter member Sir Joseph Banks, horticultural adviser to kings and longtime president of the Royal Society, the most influential scientific body in England. Sir Joseph was a mover and shaker who knew how to establish organizations on a solid footing. Knowing how the English—and none more dearly than the horticultural English—love a lord, he steered the Earl of Dartmouth into the presidential chair at the first election. Predictably, among the ninety-three new members next admitted, ten were either peers or the sons of peers. By 1809, when the membership totaled 576, including "many of the most distinguished names in the kingdom," the heady aura of nobility was such that the rank-and-file designation of "member" was upgraded to "Fellow," an appellation still employed.

In the original prospectus of the Society its objectives were defined in part: "Horticulture . . . may be divided into two distinct branches, the useful and the ornamental; the first must occupy the principle attention of the members of the Society, but the second will not be neglected." By the second decade of the century a shift in emphasis had begun. Interest in "ornamentals" (decorative trees, shrubs, and flowers) was taking over at the expense of "useful" fruits and vegetables.

There were several reasons for this. Panic over the threatened food shortages had subsided. The population was breeding merrily away at a rate of increase reaching 30 percent per decade, but improved agriculture had more than kept pace. Napoleon's fleets were no longer a blockading threat;

Nelson had demolished them at the Nile and Trafalgar. Also—and this was probably the key factor—the landed gentry had become a bit bored by scientific farming. It was still a "good thing," of course; no question about that. Overseers and estate agents must continue to keep up with the latest methods. But for the lord of the manor, increasing the size and frequency of pig litters had lost its luster. He'd accumulated wealth and wanted to spend it. What better way than in refurbishing his estate with avenues of trees and terraces of shrubs and flowers?

"Improvement" of grounds, as Lancelot ("Capability") Brown, the great landscape architect of the previous century, defined his practice, could cost a great deal of money, but as long as the Napoleonic wars and the era of industrial explosion lasted the English gentry had it. It's difficult to translate currency equivalents across a hundred and fifty years; however, in a rough approximation, the Duke of Northumberland, who drew £150,000 through farming and rents, enjoyed an income of some $3,750,000 * in modern terms. The owner of Berkeley Castle did even better: he was the recipient of approximately $4,500,000 annually.

These were exceptional holdings, of course, but lesser estates could be just as profitable in proportion. In the single county of Cheshire in the year 1812, there were no fewer than fifty properties that returned incomes ranging from $45,000 to $250,000 per annum. As for nonagricultural investments, there is the case of the Bridgewater Canal, capitalized at £200,000 ($5,000,000), and returning dividends of $2,500,000 *annually*.

To satisfy their newly acquired interest in ornamental plants, estate owners turned to English nurseries but were disappointed. The traditional range of native stock—oaks, elms, sycamores, and the like—didn't begin to meet the expansive mood of the times. What purchasers wanted were the plants they saw illustrated in the botanical magazines—camellias, azaleas, and peonies from China—or the amazing novelties brought back by the American explorers Lewis and Clark from their dash to the Pacific. Where could such plants be obtained? If not from conventional sources, then some enterprising organization must send out expeditions to find them, an organization, for instance, like the Horticultural Society of London.

The Society's council was dismayed by the clamor. A majority of the council members were conservatives, still dedicated to the "useful" cultivation concept, and were taken aback by the passions aroused by ornamentals. They tried to evade the issue as long as possible, but finally, in 1821, when the membership numbered some fifteen hundred Fellows, in-

* A multiplication factor of five (x 5) is used to compensate for the shrinkage in value of the pound sterling, and a similar factor (x 5) is employed to calculate the equivalent in dollars. Taxes, even in wartime, were trifling.

cluding the Emperor of Russia and the kings of Denmark, Bavaria, and the Netherlands, all ardent ornamentalists, they were forced into action. During the next two years expeditions were authorized and collectors sent to China, Africa, South America, and the West Indies.

For a while contentment reigned in the Fellowship as wonderful and varied flora journeyed back to England. Journals describing the specimens, and illustrating many with hand-tinted portraits, were awaited as eagerly as installments of popular novels. Seeds and roots were distributed to the landed gentry and propagated in the Society's garden.

But in time complaints began to penetrate the council chamber again. Exotics from Africa and the West Indies were all very well, but they could be grown only in hothouses and hence didn't solve the problem of estate ornamentation. What was needed was a different kind of flora—frost-proof, "hardy" plants, collected in regions of cool summers and raw winters, adaptable without coddling to the perfidious climate of Albion.

The council met joylessly in the spring of 1823 to consider the situation. It was all the more dispirited because two of the Society's collectors had died during the year, one of disease in East Africa, the other shortly after his return to England suffering from an illness contracted in China. So now an additional young man must be found ready to risk his life. Clearly he must possess faultless good health and rugged endurance—qualities most often associated with Scotsmen.

A resolution was passed instructing the secretary to write William Jackson Hooker, professor of botany at Glasgow University, inquiring if he could recommend a suitable candidate. As it happened, Hooker could—a young gardener by the name of David Douglas.

David Douglas was born in 1799 in the village of Scone, Perthshire. Once the proud site of the crowning of Scottish kings, Scone had lapsed into insignificance and disrepair by the time David appeared. John Douglas, David's father, was a stonemason, a competent but unambitious craftsman. He is pictured as a dour, withdrawn Presbyterian, possibly spending more evenings in a corner of the local tavern than at home.

David's mother seems to have been the driving force in the family. Born a Drummond, she evidently felt superior to the branch of the Douglas clan into which she had married, and constantly impressed upon her children the necessity to work their way back to respectability. David's two brothers grew up to achieve satisfactory status in clerical jobs; one sister advanced herself and the family appreciably by marrying an architect, while the other two, finding no suitors up to their mother's standards, remained single. Only David in this well-conditioned family proved recalcitrant.

He seems to have proved it early, in uncontrolled tantrums when he couldn't have his way, a line of behavior that would win little sympathy in a strict Scottish household. We find him already clamped into school by the age of three, presumably in the hope that teachers could impose discipline where parents had failed. It was a vain expectation; conflicts with the school authorities were simply added to his troubles at home.

David is supposed to have been bored by books, caring only for the lessons of nature. Tradition has it that, braving the inevitable sting of the schoolmaster's thong, he whiled away truant hours roaming the woods and the meadows, finding tongues in trees and books in the running brooks. It is an appealing picture, but we must beware of Parson Weemsisms that tend to invent anecdotes about the child to fit the character of the grown man. David later showed such capacity to study subjects of interest to him that we may at least suspect that a compelling reason for his truancy and unruly behavior at school was a rather desperate effort to call attention to himself. Unable to break through the cold disapproval of his mother, the indifference of his father, and the blindness of his teachers, it seems likely that David developed a pattern of frustration and defiant behavior that could not help evoke consequences and possibly even grudging respect.

The first glimpse of a brighter and more challenging world occurred in David's eleventh year when his father, despairing of his education, removed him from school and apprenticed him in the garden of the great neighboring estate, Scone Palace. The head gardener, William Beattie, thus became the first in a line of firm but kindly men in authority who, providing the approval and encouragement for which David had been starved as a child, were to influence him inordinately for the rest of his life.

For the next seven years, under Beattie's strict but rewarding supervision, David learned practical gardening. There were no longer disciplinary problems. The boy, delighting in the study and care of plants, exerted himself to master an exacting trade. He found that it was a trade with good employment prospects—especially for Scotsmen. Throughout the British Isles gardeners from north of the border were highly esteemed, partly because of their thorough training, but even more because of their "habits of industry, attention, and frugality" and supposed reliability of character. No doubt, when David allowed himself to dream, he saw himself as a head gardener some day—not quite as elevated as head clerk, but still, perhaps, respectable enough to win his mother's approval.

His apprenticeship completed, Douglas moved on for further seasoning to Valleyfield, the estate of Sir Robert Preston. Sir Robert and his head gardener, Alexander Stewart, were developing an extensive collection of exotics.* Douglas had never seen such displays of palms and ferns as filled

* Exotic—meaning, literally, a plant that is not native but imported—developed a secondary significance as a rare and luxurious specimen, usually difficult to grow.

the hothouses, and the profusion and variety of orchids were overwhelming.

He set himself to learn the names and origins of these intriguing novelties. Alexander Stewart, perhaps in self-defense against the ceaseless queries, obtained the run of Sir Robert's library for his inquisitive gardener, and Douglas pored over botanical texts, dictionary in hand, working his way through the Latin descriptions, learning to distinguish between the points of identification so dear to the hearts of systematic botanists. Unquestionably the young man's enthusiasm and studious application made an impression on Alexander Stewart. Among the score or so of young gardeners under his direction, all hard-working, competent and ambitious, Douglas stood out for his qualities of mind. A rapport developed between the accomplished head gardener and eager beginner that was to have far-reaching consequences, for when an opening occurred at the botanic garden of Glasgow University, Stewart recommended Douglas for the job.

Douglas reached Glasgow in the spring of 1820. It was strategic timing since, later that year, a new professor of botany, who was to shake up a good many lives, including that of the young gardener, took up residence at the university.

The professor's name was William Jackson Hooker. He was thirty-five years of age, an Englishman, good-looking, married, and reputed to be independently wealthy—alone enough to give him status in a provincial Scottish city. Professionally, he was a stimulating teacher, an energetic field botanist, and a shrewd appraiser of promising young men.

Hooker launched his regime at the university by a series of public lectures. Their success was immediate because "he was an instructor whose enthusiasm overflowed. . . . He had the art of making [his audiences] love the science he taught. So popular were his lectures that they were attended not only by students of medicine, for whom the course was primarily designed, but also by gentlemen from the city and military officers from the barracks." More than two hundred people jammed into the "small, dingy" lecture hall from eight to nine o'clock in the morning—including David Douglas, through the friendly dispensation of his new supervisor, Head Gardener Stewart Murray.

Hooker, as regius professor of botany, was also responsible for the garden, and immediately set about converting the drab eight-acre plot into a center of botanical and horticultural interest. In addition to the medicinal herbs needed for classroom demonstrations, he experimented with as many imported ornamentals—trees, shrubs, herbaceous plants—as could be coaxed into surviving Glasgow's cold and clammy winters. Douglas, who had thought that some of these exotics could be grown only in hothouses, was surprised at Stewart Murray's success in cultivating them outdoors, utilizing every last degree of warmth from a south wall or straw protection. Douglas may have proved ingenious at helping to solve some of these special garden problems, or he may have attracted Hooker's attention in the lecture hall; however it was, by early spring the professor had evidenced interest in the new gardener and invited him on botanizing trips.

Hooker was by then collecting materials for a second edition of his book on Scottish flora. He hoped to compile a text so complete that it would include every species and variety of tree, shrub, flower, grass, fern, and moss on the mainland and on the surrounding islands. It was an enormous undertaking, and Hooker was accepting all the help he could get. He corresponded with botanists from the Highlands to the Hebrides, and encouraged his students to botanize on his behalf during vacations. Inevitably, however, the major burden of the research devolved on his own shoulders.

At every opportunity, pack on back, Hooker took to the countryside. Increasingly, after the spring of 1821, Douglas became his companion. Two men, ranging at a distance from each other, eyes alert for discovery, could cover double ground. Besides, Douglas, when he became practiced at it, could take over most of the tedious work of pressing, mounting, and annotating specimens.

Anyone who has sentimentally pressed a flower or a fern between the pages of a book has unconsciously produced a botanical specimen. Naturally, for scientific purposes, rules of form and selection have been developed, but the lover's keepsake is still its basis.

An effective botanical specimen preserves, for future study, significant identifying parts of a plant. These will include, if possible, twigs, leaves, flowers, and fruits, or as many such parts as can be collected. The material is pressed and fastened (glued or tied) to a specimen sheet—a cardboard of heavy rag paper about twice the size of a page of typewriter paper. Written on each specimen sheet should be such facts as date of collection, locality where found, elevation above sea level, and other data which may be of value in future analysis.

The size of the plant collected can cause problems. In the case of a small plant—a woodland violet, for example—the entire structure, roots included, can obviously be mounted. A pine tree is more difficult, though a section of twig with needles attached, plus segments of bark and cone, together with a penciled sketch of the entire tree, should be enough to establish identification.

Under Hooker's tutelage Douglas rapidly became an expert at preparing specimens. His work, still to be seen in herberia (specimen libraries), is noted for its organization of material and clean lines. The durability of this kind of preservation is astonishing; flower colors and tones of leaves and bark are almost as vivid today as when Douglas mounted them a hundred and fifty years ago.

Though Douglas became Hooker's constant botanizing companion, we are not to suppose that Hooker was drawn to the young gardener by easy sociability. Douglas had never learned the knack of small talk and, when Hooker first took up with him, he was probably so awed by the honor that he was tongue-tied much of the time. If the professor had been looking for affable companionship, he could more readily have found it among his articulate medical students. He enjoyed being with them; every year he took the entire class into the Highlands for a week-long botanizing excursion, and many of his pupils would have been flattered to tramp around Glasgow with him as well. But Douglas was his chosen disciple—why?

Probably, in the last analysis, because he was always available when Hooker wanted him. He had no sisters and aunts to entertain over the weekend, and he could devote nights to mounting botanical specimens without the distractions of beer or cards. When vacations came and students scattered, Douglas was still available, ready to travel at a moment's notice. He was strong; he could carry the full load of their botanizing equipment and still keep up with the hot pace set by the professor. He could make dinner out of bread and cheese and accept a night's lodging in a hayloft without complaint. He could tramp all day through the rain, buffeted by the wind, and still have the enthusiasm to cheer when they found, or thought they found, a plant "new to science."

In the course of time, Douglas couldn't help thawing to Hooker's bubbling personality. He didn't become garrulous—not at that stage in his career—but he certainly found his tongue enough to qualify as an intelligent and responsive companion. No doubt the professor's attractive young family helped in the transition. Though Douglas never was invited to the Hooker home for the famous breakfasts served after lectures, he became a fixture for supper on Sunday nights, taking his turn at reading from the Bible and joining in the family prayers. The children idolized him and he made a special friend of young Joseph Dalton Hooker, one day to be a famous botanist in his own right.

When the letter from the Horticultural Society of London came in the spring of 1823, Hooker had no doubts about his recommendation. Much later he was to describe Douglas in his twenty-fourth year as possessed of "great activity, undaunted courage, singular abstemiousness and energetic zeal." In time, as the cracks developed in Douglas's personality, Hooker recognized the changes but never wavered in his steadfast loyalty and affection for his young friend.

Reciprocally, when Douglas left Glasgow, he left home in the only real sense he was ever to know a home. It was where he had found his life work, set his ambitions, and begun his uncertain maturing. He turned back to that home again and again; physically when he was able; if not, by thought, gift, and letter. There is no record of his ever having written to his natural father at Scone, while his letters to Hooker are as close to being intimate and revealing as it was ever possible for Douglas to be.

Douglas's Travels in
EASTERN
UNITED STATES
and CANADA
0
Miles
100
N
LAKE HURON
UPPER CANADA
(PROVINCE OF ONTARIO)
Kingston
LAKE ONTARIO
NIAGARA R.
MICHIGAN
ST. CLAIR R.
LAKE ST. CLAIR
Detroit
Amherstburg
LAKE ERIE
NIAGARA FALLS
Buffalo
ERIE CANAL (UNDER CONSTRUCTION)
Rochester
ERIE CANAL (COMPLETED)
MOHAWK R.
NEW YORK
Albany
Clermont
HUDSON R.
VERMONT
N. H
MASS.
CONN.
New York
Hoboken
Flushing
LONG ISLAND
PENNSYLVANIA
DELAWARE R.
Bordentown
Philadelphia
NEW JERSEY
MARYLAND
VIRGINIA
ATLANTIC OCEAN

2

COUNTRY OF THE NEW PEOPLE

THE LAST DECADES OF THE EIGHTEENTH century developed into a totally unexpected Age of Discovery. It had been thought that three hundred years after Columbus, da Gama, and Magellan, there was little of consequence left to be found in the world. Yet during a thirty-year period, beginning in 1770, a continent (Australia *), an important archipelago (Hawaiian Islands), and a major river (the Columbia) turned up in rapid succession.

Botanists on the discovery ships, or on ships following soon after discovery, collected, pressed, and mounted specimens of the new flora, but they rarely attempted to take back living roots or seeds. That was grubby work to be performed later by plant collectors. There was usually a lag between the visit of the botanist and the arrival of the plant collector. For one thing, it took time to build up a demand for novel shrubs and flowers; for another, conditions in the new territory had to be reasonably stable for the collector to work effectively. A botanist might climb back aboard his ship with an arrow through his hat and be none the worse for his adventure. But a plant collector, who must patiently wait out the seasons, locating species in the spring and summer, returning in fall and winter to harvest seeds and dig up roots, needed a settled community to serve as his base.

* Mariners had sighted dry and forbidding shores a couple of times in the previous century, but it wasn't until Captain James Cook touched the fertile east coast in 1770 that the exploration of Australia really began.

The Horticultural Society had hoped to send David Douglas to the Pacific Northwest under the protection of the Hudson's Bay Company, the great fur monopoly whose empire extended across the entire American continent. It was expected that he would travel up and down the Columbia River, enjoying the hospitality of the Company's trading posts while collecting the plants made famous by the Lewis and Clark expedition of 1804–6. Unfortunately, while Douglas was in transit from Glasgow to London, word was received from the Hudson's Bay officials asking for a postponement of his mission. Organizational difficulties were given as the reason, a situation that would right itself within a year or two.

Next in order of preference as a plant-collecting arena was the north coast of China. English gardeners were already familiar with the spectacular flora of the south coast, since the East India Company had been trading out of Canton for years, but many of the plants were too tender for the British climate. It was expected that farther up the continent, at about the latitude of Shanghai, hardier shrubs and flowers might be obtained.

Much hope had been invested in a recent ambassadorial mission sent to Peking to petition His Celestial Highness to open up northern ports to British shipping. But a hitch had occurred. The British ambassador, refusing to grovel before the emperor, was denied audience until he did, thus bringing negotiations to a standstill. All in all, it seemed doubtful that Douglas would travel to north China in the foreseeable future.

Finally there was Chile. Little was known about it, but it presented possibilities. Stretching as it did from the South Polar seas to well above the Tropic of Capricorn, it must surely contain somewhere within its eellike contours a climate comparable to England's, if not at sea level, then higher up in the Andes. Baron Alexander von Humboldt's theory that climate depended as much on height above sea level as distance from the equator had gained wide circulation in British scientific circles. But David Douglas was not to put von Humboldt's theory to the test in 1823. Simon Bolivar and like-minded liberators were making the climate uniformly hot everywhere in Chile that year.

The truth is that when Douglas reached London, ready to risk life and limb if need be to prove his value to the Horticultural Society, his new employers had no place to send him. No place, that is, until a destination was hastily cooked up, passed through the council, and announced to the membership. Eyebrows must have lifted in the palatial halls of England when it was learned that the new collector was to be dispatched to New York, of all places, and not to beat the bushes for ornamentals, either—though what could still be novel in Yankeeland after a century of intensive botanizing was hard to imagine—but to investigate commercial fruit trees and purchase a collection of the newest and best varieties.

That objective, fruit trees, must have been a tip-off for the more cynical Fellows. Only one man could have both conceived that mission and pushed it through the council so rapidly, a man with a special interest in improved fruit varieties. His name was Thomas Andrew Knight and he cultivated extensive commercial orchards in Shropshire. Twelve years earlier he had been voted into the president's chair, succeeding the Earl of Dartmouth.

It had seemed a worthy succession at the time. Knight, though technically a commoner, could trace his ancestry back to the Norman conquest. He lived in a moated, turreted pile known as Downton Castle and owned more arable, income-producing land than half the nobility of England. He had seemed thoroughly solid; yet now many Fellows began to wonder. There was something underhanded about committing Society funds to fruit trees when ornamentals were the order of the day. Not quite *noblesse oblige*, to put it charitably. Perhaps after all, and despite the extenuating circumstances, it had been a mistake to elevate a commoner to the presidency.

It was a mistake rarely to be repeated. Of the twelve presidents who have succeeded Thomas Andrew Knight since 1838, one has been a prince—Albert, consort to Queen Victoria, during whose regime the name was changed to the *Royal* Horticultural Society; two were dukes; six, assorted lords, ranging from earls to viscounts; and two were knights, leaving only one commoner to climb again to the highest rung. Though there have been periods of mismanagement as well as scandals and debacles during this aristocratic epoch, Fellows have always had the satisfaction of knowing that mistakes have resulted from honest stupidity; never again has there been a suggestion of anything underhanded in the Royal Horticultural Society.

To David Douglas, riding outside on the Scottish Mail, arrival in London must have been both an exhilarating and a bewildering experience. London, with a population already in excess of a million and a quarter, was, then as now, Rome, Jerusalem, and Mecca, the spiritual and temporal center of the British universe, a metropolis compared to which every other city in the land was dull and provincial. Through the illustrated magazines then becoming popular, Douglas may have expected some of the sights that greeted him—wide streets with avenues of trees, elegant buildings, fashionably dressed dandies and their ladies—but nothing could have prepared him for the hullabaloo and uproar in which his journey ended.

As the coach rolled into the terminal, horses steaming, a wave of people swept in from all sides. There were hostlers in enormous boots, greeters and criers from inns and lodging houses, hawkers of all kinds, hustlers of both sexes, and just plain friends and relatives of incoming passengers. Hysteria always seemed to grip this motley throng, commentators of the time agree, causing individuals to scream at the top of their lungs and fling themselves about as if in the throes of religious ecstasy.

Somewhere in that crowd, probably holding back until the first shock waves had broken, was an emissary from the Horticultural Society. Possibly it was William Christie, foreman at the Society's garden in Chiswick. Christie was, of course, a Scot, and it would be comforting for Douglas to hear a familiar burr again after days of coping with high-pitched English voices.

The men would travel toward Chiswick in a horse-drawn cart, with Douglas's luggage stowed under the plank serving as seat. Douglas would be curious about everything connected with the garden and would keep Christie busy answering questions. Though the thirty-three-acre plot had only been acquired the previous year, twelve hundred varieties of roses were already in the ground. In addition, an experimental kitchen garden had been started and some three thousand fruit trees planted. The collection of ornamentals was still small but growing.

Christie would have nothing but praise for Donald Munro, the head gardener. He might also sketch the characters of the younger gardeners with whom Douglas was to share quarters. But it's doubtful if he presumed to offer opinions on the garden's administrative personnel. Silence was the safer course. Let Douglas find out for himself some of the peculiar things that went on.

Douglas found out soon enough. Take the matter of labels. He was used to meticulous labeling procedures at Glasgow, where neat markers spelled out both the botanical and common names of each plant for the benefit of students and visitors. At Chiswick there were no such labels. When Douglas inquired why, he was told that Mr. Sabine didn't believe in them.

Joseph Sabine, the honorary secretary of the Horticultural Society, ran Chiswick garden with a firm—some said dictatorial—hand. He didn't believe in labels because there was a purpose to be served in omitting them. When Fellows visited the garden, members of the staff, thoroughly drilled in nomenclature, were assigned to accompany them about the grounds, identifying plants on request. Afterward a report was filed, detailing all questions asked and all comments made during the tour.

The system was wasteful of the employees' time, but it provided Sabine (rhymes with "cabin") with a running record of what was thought and said about his administration. He escorted distinguished visitors and titled Fellows about the grounds himself, but he never put much reliance on comments made to his face.

There was always something a little devious about Joseph Sabine. Perhaps it wasn't altogether his fault. In an age when wealth and position were everything, he was handicapped by having been born into a poor if respectable family. As a result he had been forced to make his way by a combination of ability and guile. Trained as a lawyer, he had pursued an undistinguished career at the bar until, approaching forty years of age, he wangled an appointment as inspector general of assessed taxes. Just what he assessed, or when, was never quite clear, but, with an assured income of £1,200 per annum (the equivalent of some $30,000 today), he was finally able to cut a figure as a scientific man-about-town. In short order he became a member of the Royal Society, the Linnean Society, and the Zoological Society. But it was in the Horticultural Society that he was to create a stir long to be remembered. Joining in 1810 he had, by 1816, maneuvered his way into the influential post of honorary secretary.

Over the next fourteen years Sabine was to transform this office into a command post of English botany. As editor of the Society's journal—*Transactions* as it was then arcanely called—he controlled the publication of articles upon which botanical reputations could be built. As administrator of the Chiswick garden, he had sole responsibility for distributing rare seeds, roots, and plants to Fellows, nurserymen, and overseas correspondents. Under the circumstances, many prominent people found it useful to court the secretary with flattery, gifts, and like blandishments.

Sabine lost no time in beginning Douglas's indoctrination. Books were assigned and long lists of grafted fruits and cultivars ordered memorized. Instructions in Chiswick's routines and methods were left to Head Gardener Munro, but, in the final weeks before Douglas's sailing, Sabine took over again for an intensive course on social deportment in the United States.

Like most conservative Englishmen, Sabine still distrusted the institutions and philosophies of the new republic. He didn't go quite as far as Dr. Johnson who, some fifty years earlier, had advocated anything "short of hanging" for the new nation's churlish citizens, but he deplored the egalitarian notions encouraged by democracy and did his best to protect Douglas from being influenced by them.

He impressed upon the young Scot that his trip to the United States was only a preliminary test in anticipation of a greater expedition. But it was to be a searching test, especially in regard to his conduct as a representative of the Horticultural Society. He was not to put on airs or exaggerate his importance among strangers. He must remember that his mission was to buy fruit trees, and this he must do expeditiously, with as little cost to the Society as possible.

In his spare time and during his cross-country trips he would be expected to botanize the native flora and return with pressed specimens of plants he collected. He would also be required to keep a journal, recording daily activities, which was to be delivered to Sabine on his return to England.

There is no doubt that before the indoctrination was over, Douglas was in awe of the honorary secretary; no doubt, either, that Sabine took advantage of the young man's inexperience to transfer his loyalty from the Society to himself. In future, Douglas's hopes for recognition, advancement, even pocket money, were to hang on what Sabine could get for him.

The sense of dependence was strong enough for Douglas, ten years later and half a world away, to compromise his career through mistaken loyalty to a man who had never been truly his friend. The irony is that Sabine then no longer cared. He had tumbled so far from his former glory that no gesture of allegiance could restore him.

Douglas sailed from Liverpool early in June, 1823, and reached New York two months later. His accommodations were probably well below decks since it wasn't the policy of the Horticultural Society to coddle its collectors. The guidelines on the matter had been laid down by the great Sir Joseph Banks. As honorary director of His Majesty's garden at Kew, Banks was an old hand at dispatching botanical travelers to foreign parts, and held precise opinions as to their proper rank. "Collectors must be directed by their instructions not to take upon themselves the character of gentlemen," he ruled, "but to establish themselves in point of board and lodging as servants ought to do."

New York was, even then, a not inconsequential city of 125,000 population. It was already established as the premier port of the Atlantic coast, and drew its wealth mainly from shipping and mercantile activities. A prosperous middle class had grown up, and their fine houses extended north to the newly developed Washington Square area. Beyond that, fields, woodlands, and scattered villages extended to the upper end of Manhattan Island.

Douglas, carrying letters of introduction from Sabine, began calling on New Yorkers who were overseas (or "corresponding") Fellows of the Society. An early visit was to Dr. David Hosack, a prominent physician. Hosack insisted that the young collector stay for dinner, urged that his home be considered the Society's headquarters for messages, meetings, and any other business, and arranged for a guide to introduce Douglas to the sights of the city.

Escorted by his guide, Douglas visited the Fulton produce market. He admired the beets, carrots, and onions, but thought the peaches looked "sickly" and the cauliflower "miserably poor." He visited nurseries in the city, examining the latest developments in fruit trees, then journeyed up Manhattan to see the Elgin Gardens. This was the first botanical garden in New York, established by Dr. Hosack and maintained by him while he waited for the city fathers to take it off his hands as an educational institution. But they never did. The gardens were considered too far out of town and, besides, the citizenry showed little interest in plants that did not produce fruits, could not be eaten, or were not convertible otherwise into cash. Eventually Dr. Hosack was forced to sell the land. He'd have done his descendants a favor by holding on because the site of the Elgin Gardens is where Rockefeller Center stands today.

But Dr. Hosack was to receive an honor that must have pleased him more than the prospect of making his descendants rich. Within a few years, David Douglas was to discover on the Columbia River a new genus of the pea family which he named *Hosackia* after his New York host. To a botanist —and Hosack had been a professor of botany at Columbia College—such recognition amounted to immortality, placing him in the select company of Father Camellus, William Forsyth, and Dr. Clarke Abel, names forever blended into the glories of *camellias*, *forsythias* and *abelias*.

In mid-August, Douglas traveled by steamboat and stagecoach to Philadelphia, primarily to visit nurseries, but also to see for himself the plant introductions of Lewis and Clark, now growing in various gardens of the city. He also hoped to see the pressed and mounted specimens collected by Meriwether Lewis during the Columbia Expedition of 1804–6, but he found few in evidence when he visited the University of Pennsylvania. The bulk of that irreplaceable collection had been spirited away through neglect and larceny.

Benjamin Smith Barton, professor of botany at the university, had provided the neglect in generous quantities. It was to Barton that Thomas Jefferson entrusted the amazing flora of the Columbia River for classification and description. Incredibly, Barton postponed the work for years. When he finally got around to it, it was too late. Most of the collection had been purloined by his former assistant, Frederick Pursh, who had long since departed for his native Germany. With the publication of *Flora Americae Septentrionalis*, a work largely based on the Columbia material, Pursh secured for himself the international botanical reputation Benjamin Smith Barton might have had—and also, perhaps, a niche in the pantheon of infamy.

Returning to New York by way of the Delaware River, Douglas stopped off to visit an estate in Bordentown that surprised him with its elegance despite its location in the backwoods of New Jersey. "A most splendid mansion," he noted in his journal. "Fields well cultivated, pleasure grounds laid out in the English style; there were many fine views." He revealed in some surprise, "Here stands the house of Joseph Bonaparte."

The sometime King of Naples and dethroned monarch of Spain was

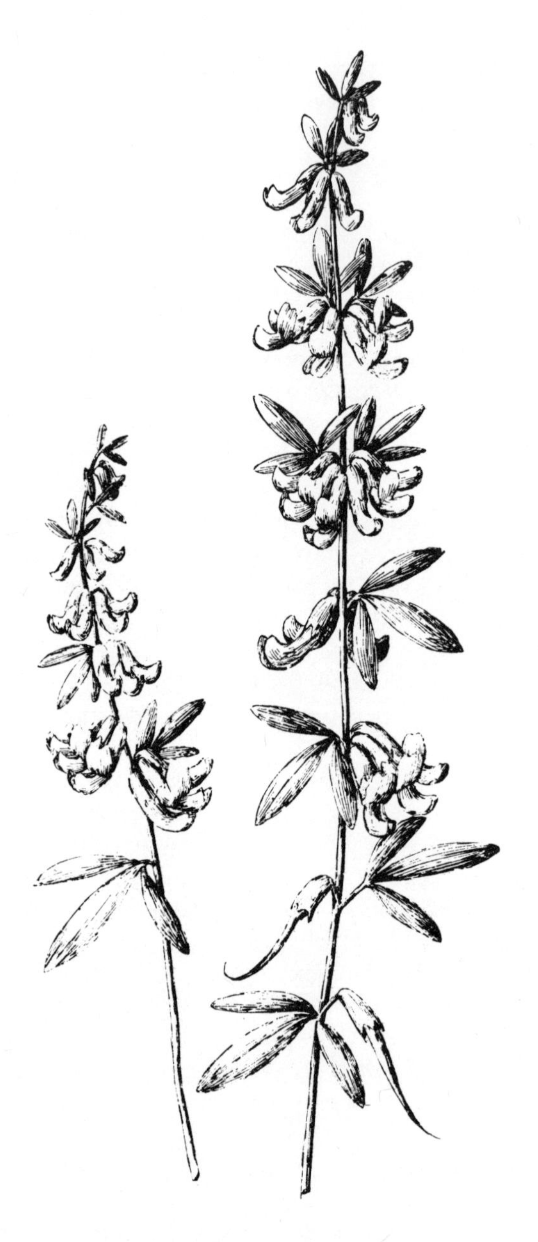

HOSACKIA BICOLOR (*bird's-foot trefoil*)

A plant genus, new to science, that Douglas discovered in the Pacific Northwest and named for his former host in New York City, Dr. David Hosack, a founder of Bellevue Hospital and the attending physician at the Burr-Hamilton duel in 1804.

indeed domiciled in New Jersey. Though neither he nor his silk-coated courtiers fit the description of "wretched refuse of teeming shores"—a class of immigrant to be welcomed by the Statue of Liberty in later years—they certainly yearned to breathe free and, after Waterloo, the unpolluted atmosphere of Bordentown had its attractions.

According to Joseph Bonaparte's own account, only last-minute scruples prevented his celebrated brother from becoming still another Jerseyite. Napoleon had ordered the escape but, just before embarking, he paused to consider his legend. He decided that history would deal harshly with the First Citizen of Europe if he saved himself while his country lay prostrate, so, instead, he surrendered to the British.

We are not to suppose that it was all *joie de vivre* for Joseph and his entourage in New Jersey. For one thing, there was boredom. Until 1821 they were kept busy plotting Napoleon's escape from Saint Helena, but his death robbed them of further occupation in that direction. For a while they contemplated creating an émigré colony in upstate New York, but there were too many trees to cut down. Finally, after twenty years of exile, Joseph threw in the sponge. Tired of American crudities—he'd had to make do with a Quaker mistress—he returned to Europe to die with verve and style in Florence.

In September, Douglas left New York City and began a journey northward into Canada. This extension of his trip had been authorized in order to test his botanizing ability in the comparatively unexplored wilderness of what was then called Upper Canada, the modern Province of Ontario.

He stopped off in Albany to pay his respects to DeWitt Clinton, several times governor of the state of New York and an overseas Fellow of the Horticultural Society. Clinton was then involved in pushing to completion the construction of the Erie Canal, the waterway that was to connect the Great Lakes with New York Harbor. Douglas presented his letter of introduction at the governor's office and was politely withdrawing when the great man came rushing after him. He insisted that Douglas stay for lunch and, before he let him go, pressed into his hand introductions to friends along his route to Canada.

From Albany, Douglas traveled west by stagecoach over corduroy roads, then by barge on newly opened sections of the new canal, and finally by steamer across Lake Erie to Amherstburg, a garrison town and trading post across the river from Detroit, Michigan, then a hamlet of some fifteen hundred people.

Douglas's host in Amherstburg was Henry Briscoe, an officer in the

British garrison and a friend of the Sabine family. Douglas noted in his journal that "I received great kindness" from Briscoe, but tactfully omitted mention of the probable circumstance that obliged him to lodge in town rather than at his host's house. Briscoe, it can be assumed, was married to an Indian and, as was often the case in such unions, took pains to keep his wife and children out of sight.

Later, on the Columbia, Douglas was to become increasingly familiar with the secretive, whispering world of mixed alliances, but encountering it for the first time, and probably without preparation, could have been something of a shock. "Mr. Briscoe . . . readily pledged his exertions for the furtherance of the Society's objectives," he wrote in his journal somewhat irrelevantly, as if not knowing what else to say about his host at their first meeting.

Briscoe's "exertions" seem to have begun and ended when he accompanied Douglas into the woods on the first fine day of his visit. He took his gun and his dogs along so he could kill birds when the botanizing became boring. Despite the distractions, Douglas reveled in the immense oaks, walnuts, and beeches of the virgin hardwood forest, "woods of astonishing magnitude," he called them. In the clearings, when the sun penetrated through the leafy canopy, he "had an opportunity of seeing in perfection" sunflowers, asters, and goldenrod.

On succeeding days, Douglas botanized alone; in the woods, on an island in the Detroit River, and later on a trip to Lake Saint Clair. He was in an area particularly rich in varieties of oaks, always a tree of powerful emotional appeal to the British aristocracy, suggesting stout ships,* bristling with cannons to defend England. Douglas diligently collected acorns, enough to guarantee an invincible oaken fleet for England in about one hundred and fifty years (say, 1973). It may have been in the woods of Amherstburg that he developed his technique of shooting down leaves and fruits from trees too tall to climb, a practice that was later to prove invaluable on the Columbia.

Near Lake Saint Clair a bizarre incident occurred. Douglas hired a cart and driver to guide him through the unfamiliar district. The driver was black, one of many fugitives from the Southern states who had been helped to freedom in Canada. Some found jobs as field hands; others, like Douglas's driver, preferred less-regimented labor.

* The statistics of oak consumption in building men-of-war were truly staggering. One medium-sized battleship mounting seventy-four guns (Nelson's *Victory* was a hundred-gunner) required for construction two thousand trees of at least a hundred years' growth, each tree calculated to yield only a ton of prime timber.

The day was hot and Douglas, interested in a clump of mistletoe, removed his coat in order to climb to it. "I had not been above five minutes up," he tells us, "when to my surprise the man . . . took up my coat and made off as fast as he could run with it." Douglas gave chase but the thief escaped, carrying off money, field notes, and a textbook. Seething with indignation, Douglas found his way back to town, but received little satisfaction from the owner of the cart, who refused to accept responsibility for his employee.

As far as we know, Douglas encountered Negroes only twice in his life —on this occasion and in Hawaii, when the Reverend Diell's John was forced upon him as a companion. Each encounter is curiously disjointed, unhinged to any event that preceded or followed it. If one is to look for significance— and there is no real reason to—the incidents might be taken as an indication of chaotic forces and conflicts already building up outside the neat and ordered world of the nineteenth century.

Despite mishaps and poor weather, Douglas continued to botanize, collecting coreopsis, lobelias, and shellflowers from damp locations; phlox, gayfeathers, and lady's slippers from peaty soils along the river. He obtained seeds where possible; where not, he dug up entire plants, hoping to keep the roots alive until he reached England.

The sheer bulk and height of the surrounding trees continued to impress him. It was a "climax" forest of hardwoods. Over thousands of years the woods had matured through stages, first shading out the fast-growing birches and poplars, then eliminating the conifers, until finally the ultimate masterpiece was achieved, a forest composed entirely of oaks, maples, walnuts, beeches, ash, and related timbers. It was a rare phenomenon and Douglas knew it. "The woods were of astonishing magnitude," he tells us; and again, "There are no pines here. . . . [But] trees from fifty to seventy feet high, forty feet without branches."

But destruction had already begun. Douglas describes a district called the French Settlement which, through clearing the land for orchards of apples, pears, and peaches, was winning a reputation as the Eden of Upper Canada. He saw oaks frequently measuring "from 20 to 25 feet in circumference . . . from 80 to 100 feet high," felled and, for want of sawmills or markets for lumber, burned where they lay. Ashes piled up like dirty snow drifts across the land, some of it sold to process lye.

Still, the inroads that men with axes and torches were making into the woods in 1823 appeared slight. The great forests stretched to limitless horizons and it seemed impossible that they could ever disappear.

Late in September, Briscoe "and family" were ordered to move to

Kingston on Lake Ontario, another garrison town. Since the weather had turned cold, Douglas was ready enough to pack his collections and head back for the United States. The steamboat took him to Buffalo, and from there he made his way overland down the Niagara River to the falls. Douglas dutifully recorded in his journal that he was "impressed with their grandeur," but his business was plant-collecting, and soon he was scrambling among the rocks investigating clumps of green in the mist and spray. He found a vetch and a viola, "both in seed," he recorded with satisfaction.

Douglas slowly worked his way down New York State, making use of Governor Clinton's letters of introduction along the way. When he reached Albany he found the town "all in an uproar," guns firing, bands playing, to celebrate the formal opening of the eastern section of the Erie Canal. Again Douglas hardly expected Clinton, the hero of the hour, to have time for him, but again he was mistaken. The governor asked the collector to stay in town for a few days since there were still distinguished people whom he wanted Douglas to meet.

One such notable was Stephen Van Rensselaer, eighth patroon of a vast Hudson Valley land grant. Armed with a letter, Douglas arrived at the Van Rensselaer estate early in the morning expecting to be shown around by a gardener. Instead, he was greeted by the patroon himself, taken on an extensive tour of the grounds, and invited to join the family for breakfast. Douglas was amazed at this display of democratic hospitality from the man he called "the most wealthy . . . in the United States."

Van Rensselaer made no such claim for himself, and, possibly, in liquid assets, he was surpassed by John Jacob Astor and one or two other rising entrepreneurs; but no one in the young Republic could vie with him in length of ancestral line, which went back to the earliest Dutch colonization. A graduate of Harvard, Van Rensselaer had served his country as a soldier and a congressman, but he was most admired for the qualities that had surprised Douglas—genial manners and democratic behavior.

Not the least of Douglas's pleasures in Albany was running into Dr. Hosack, who had come up for the canal celebrations. Their conversation turned to botanizing, and the doctor promised, on their return to New York, to take Douglas collecting in the wilderness around Hoboken, New Jersey, where rhododendrons grew "fully 17 or 20 feet high," as Douglas was to report later, and "*Kalmia latifolia* [was] also vigorous."

In an age of fully committed men, Dr. Hosack stood out for the range of his activities. A graduate of Princeton, he had studied medicine in Edinburgh and London—in fact, had read a paper before the Royal Society with the great Sir Joseph Banks presiding. Returning to New York, he had lectured on botany at Columbia while establishing a medical practice. Not content with rising to prominence as one of the leading physicians of the city, he had fostered clinics to raise standards of medical training, and had played a leading part in founding Bellevue Hospital. He had also (though he hardly considered it his finest hour) served as attending physician at the Hamilton-Burr duel.

On October 11, Douglas left Albany with still one more letter of introduction, to General Morgan Lewis of Clermont. Traveling downriver by steamboat, Douglas reached the general's home that night, and next morning, a Sunday, attended church with the family. During the service, rheumatism gripped him in the knees. Barely able to fight off the pain, he hobbled back to his host's house and collapsed.

The Lewis family took care of him with cheerful efficiency, applying poultices, stilling his embarrassed apologies. "I can never forget the attention paid to me by General Lewis and his family," Douglas gratefully recorded. When he was able to "crawl about a little" he was driven over the farm on a tour of inspection. He seemed surprised that the general, who had only taken up agriculture at the age of sixty-nine, "has all the newest modes of tillage and many of the newest and most improved implements." He would have been less surprised if he'd known his host's long history of excelling at everything to which he turned his mind and energies.

Morgan Lewis served through the Revolutionary War and marched in George Washington's inaugural parade. Turning to the peacetime study of the law, he became chief justice of the New York Supreme Court, resigning later to enter politics and to successfully run for governor. At the time of Douglas's visit, he was combining scientific farming with biblical scholarship, and had begun the study of Hebrew so that he might read the Old Testament in the original text.

The fact is that Douglas had been meeting an extraordinary group of men since his arrival in the United States. Despite Secretary Sabine's warning against being taken in by Americans, he sensed in such people as Governor Clinton and Dr. Hosack a spontaneity of thought and action utterly impossible for their rank-bound counterparts in England. He was among new people, products of a new time and a new society, men who almost literally exploded in every direction in which their interests led them. Well born, well educated, intensely alive, they felt themselves the equals of kings, while, at the same time, no better than any citizen who could contribute to the general welfare of the country.

It was this second attitude that was difficult for Douglas to grasp. Nothing in his experience had prepared him for a society based on worth, not class, and for some time he kept pushing away evidence of its existence. He chose to interpret friendly feelings aroused by his own worth as nothing more than gestures of goodwill toward the Horticultural Society. Typically, after a visit to Zaccheus Collins, a Philadelphia botanist, he entered in his journal: "I am truly obliged to this gentleman for his kind attention and his willingness to forward the views of the Society." When Martin Hoffman, a leading Manhattan merchant, invited him to dinner and to spend the night, Douglas thought he wanted to give advice on purchasing fruit trees. It didn't occur to him that Hoffman's marriageable daughters had anything to do with the invitation.

A change in attitude began shortly after Douglas's trip to Canada. Perhaps Briscoe's scant hospitality—in fact, almost boorish behavior—in Amherstburg contrasted too sharply with the attention he had received everywhere in the States; or the secretive, purdahlike atmosphere of Briscoe's household might have come as a shock after the affectionate mingling of American families. Whatever the triggering causes, Douglas recrossed the border with a new enthusiasm for the Republic and its citizens.

The change is soon apparent in his journal. The style becomes more relaxed. Awkward constructions, such as "I received great attention from Mrs. Thomson, which was thankfully acknowledged," disappear in favor of such brisk phrasing as "I must observe that Mr. Prince Jun., did not receive us with kindness but the reverse." He tries his hand at *bon mots*—"Dr. Hosack is a Clinton, and Clinton a Hosack"—and even has a fling at humor, never his strong suit: "On asking any person if (fruits) are good, they invariably say they are not so fine as last year, and I think had I been there then, they would not have been so fine as the year before."

Nothing could better illustrate Douglas's new freedom and growing self-confidence than his clash with the Princes. These men, father and son, operated a commercial nursery in Flushing somewhat grandiloquently named the Linnean Botanic Garden. Since they were overseas Fellows of the Horticultural Society they had expected to sell Douglas the bulk of the material he had come to the United States to collect, and were outraged to hear that he was shopping around purchasing fruit trees from Philadelphia to the Canadian border. When Douglas visited them again on his return to New York, they were primed and waiting for him.

They tried, or Douglas thought they tried, to make up for lost sales by raising the price on every item in which he expressed interest. When Douglas refused to buy in the face of this profiteering, William Prince, Sr., threatened to write Secretary Sabine complaining about his cheap-jack collector. Douglas retaliated by threatening to expose the unkempt state of the Princes' nursery which, he derisively observed, "is covered over with weeds."

Fortunately the situation was eased, each side simmering down sufficiently to permit business to be transacted; but nerves were left quivering, as witness the journal entry with which Douglas dismissed the Princes: "I am sorry to say that I must leave America without having good feeling toward *every* person."

Though Douglas didn't care to mention it, he knew there was at least one other person in New York who harbored less than "good feeling" toward *him*. This was Dr. John Torrey, a physician and botanist who was even then developing a reputation as the arbiter of American flora. He was a correspondent of Professor Hooker in Glasgow; in fact, it was on Hooker's behalf—to deliver a package of seeds—that Douglas had called on Dr. Torrey soon after his arrival in New York in August.

Something happened during that visit, something dismaying to Douglas. He recorded in his journal, "I found an intelligent botanist, an agreeable person, and much disposed to aid me." But despite this apparently encouraging start, the name of Torrey drops from the journal with a finality which makes us suspect that the entry was more for Joseph Sabine's benefit than in the interest of unvarnished truth.

The truth—or Torrey's version of it—emerged later when he wrote to Hooker about the package of seeds. There were some technical matters he wanted the professor to clear up. He said he'd asked Douglas about them in New York, but had been dissatisfied with the answers, commenting, "He is such a liar I know not whether to believe him or not."

Whatever the nature of the exchange that day at Torrey's house, Douglas was clearly frightened at having roused the doctor's ire. Quarreling with nurserymen was one thing, but offending a botanist of Torrey's standing was quite another. He must have been on tenterhooks for months, fearing that a letter of complaint might be sent to Sabine, but apparently Torrey never bothered.

It is almost certain that Douglas never heard from Hooker about the matter. That amiable professor made a point of never passing along criticism that could wound or discourage, and certainly not from Torrey. He was well aware that the peppery New York botanist had at one time or another succeeded in insulting many men of science, including Hooker himself.

There is one more event to record before the brig *Nimrod*, with Douglas's fruit trees in the hold and his pigeons, ducks, and quails (gifts to the Zoological Society from Governor Clinton) caged on deck, picks its way through the winds of New York Harbor. It took place in Philadelphia when Douglas, making a last round to collect plants, was introduced to the botanist Thomas Nuttall. Their paths were to cross again at a time of anguish.

Nuttall, a quiet man of thirty-seven, already famous as an exploring botanist through epic trips up the Missouri and Arkansas rivers, and who had recently been appointed the first professor of natural history at Harvard College, seems to have taken an immediate liking to Douglas. When he found out that the young man hadn't yet visited Bartram's famous garden in nearby Kingsessing, the first botanic garden in America, he offered to accompany him. They talked botany without letup as they walked along, Nuttall flattering his companion by discussing quite technical aspects of plant nomenclature.

Like Douglas, Nuttall had been born in the British Isles—Yorkshire—into a working-class family. He was apprenticed to a printer, but, at an early age, through rambles in the countryside, discovered a passion for botany that he couldn't suppress.

Despairing of a future in England without influence or education, Nuttall had emigrated to the United States when he was twenty-two, choosing Philadelphia as a port of entry because that was where another printer, Benjamin Franklin, had been successful. He earned his living at the presses during the winter, saving enough to wander across the country botanizing when the days lengthened. He'd set up much of the type for his first book himself in order to reduce publishing costs. That book, *The Genera of North American Plants*, broke a long tradition by giving descriptions of species in English rather than in the classic Latin. The innovation shocked conventional botanists, but, since only Nuttall had collected west of the Mississippi, they bought his book anyway.

Nuttall never minded shocking people, or perhaps he didn't notice. He quietly went his way, traveling to remote outposts, collecting specimens, taking notes. Within a few years he was to resign his post at Harvard, with its then fantastic salary of a thousand dollars a year, because classrooms shut him away from the prairies and mountains where his true work lay. Nuttall's objectives in life were as direct and simple as the man himself. "To converse, as it were, with nature," he wrote, "to admire the wisdom and beauty of creation, has ever been, and I hope ever will be, to me a favorite pursuit."

Despite the difference in their ages and temperaments and outlooks, the men enjoyed a close and stimulating companionship for several days. It was one of the rare periods when Nuttall, usually silent and withdrawn, felt the need to talk. "I found [him] very communicative," Douglas tells us, delighted at his good fortune to be on hand. "Mr. Nuttall says . . . ," "Mr.

Nuttall describes . . . ," "Mr. Nuttall showed me . . . ," pepper his entries in the journal from November 1 to 3, finishing with the discovery of a rare fern four miles from Philadelphia.

They parted with genuine regret. Douglas continued to admire Nuttall and pride himself on their friendship until, years later, his regard was overcome by jealousy. When the break finally came and his regard soured, he hardly seemed aware of what had gone wrong. But perhaps it was impossible for Douglas, who wanted so much, ever to understand a man like Nuttall, who wanted so little and yet achieved everything.

3

PURPOSEFUL JOURNEY

BACK IN LONDON, ONCE AGAIN AT THE gardeners' lodge in Chiswick, Douglas must have been pleased but mildly surprised at the praise heaped upon his recent expedition. "This mission was executed by Mr. Douglas with a success beyond expectation," ran the tribute in the Society's official publication, almost certainly written by Joseph Sabine. "He obtained many plants which were much wanted, and greatly increased our collection of fruit trees by the acquisition of several sorts only known to us by name."

Unaware of the background, Douglas couldn't know that Sabine was trying to appease influential Fellows still disgruntled by the "fruit trip." The secretary sweetened his message still further by discreetly pointing out that the trip hadn't cost much: "the *presents* of cultivated plants . . . embraced nearly everything which it was desirous to obtain."

Detailed descriptions were given of twenty-one varieties of peaches and ten varieties of apples. Feature billing was also given to a pear tree said to have been imported from Holland a hundred and seventy years before and planted by Peter Stuyvesant.

Besides fruits and vegetables, Douglas had returned with an assortment of living roots, seeds, and cuttings gathered in woods and along streams from Philadelphia to Amherstburg, together with dozens of pressed and dried specimens for the herbarium. He may have been disappointed that so little attention was paid to this material but, as was to be expected, he'd not brought back much that was new from the heavily botanized Atlantic Coast. A yellow-leafed honeysuckle aroused some interest. It had already been collected, described, and named *(Lonicera hirsuta)*, but it was first grown in England from the living roots brought back by Douglas. Hence he gets credit for introducing the plant to cultivation.

As a horticultural collector, Douglas was mainly concerned with "introducing" plants already discovered by others. But often, working through the virgin flora of California and the Pacific Northwest, he simultaneously discovered and introduced plants. The sugar pine, digger pine, and red-osier dogwood are such examples, as is the trefoil that he named after Dr. Hosack of New York. In almost every category of plants to be found between Alaska and the Mexican border there is at least one species named for Douglas, either in honor of his first discovery or in respect for his collecting achievements. *Aster Douglasii, Baccharis Douglasii, Chrysanthemum Douglasii, Draba Douglasii, Eschscholtzia Douglasii*—the names march on through the alphabet. Few men had a quicker eye in detecting a novel plant in the wilds, or estimating its potential for the garden.

The term "botanist" has sometimes been loosely applied to Douglas. This is an error. Douglas was aware, and his professional contemporaries were even more aware, that he had neither the scientific nor the academic background to permit him to qualify as a botanist. Neither, for that matter, had Thomas Nuttall, in the beginning, but he had crossed the ocean to find what education he needed to make him preeminent in his field. A few Englishmen of modest birth, notably John Lindley, stayed at home and still managed to circumvent class barriers and enter the botanical establishment, but they achieved success as much through fortunate patronage as by brilliant intellects.

In this respect Douglas was not so lucky; on the other hand, his talents did not draw him toward the academic. His dedication to botany was as a practical explorer, a field specialist who sought out and brought back for study and cultivation plants never known before in Europe. And in this work he was supremely successful. What he missed in being excluded from the inner academic circle was the dignity of a title, and the lengths to which jealous professionalism went to exclude him from such honors seems to us now almost ludicrous. For instance, an officer of the Royal Horticultural Society, writing a tribute to Douglas almost a century after his death—when raising him to the full botanical priesthood could have done little harm—can bring himself no closer than to describe his subject as "this eminent Horticulturist." It was left for some uninitiated Scotsmen, erecting a monument to his memory in Scone, to commit the ultimate gaffe of calling Douglas "this eminent botanist."

ESCHSCHOLTZIA CALIFORNICA (*California poppy*)

First described by the botanist Adelbert von Chamisso in 1816 and named in honor of his friend Johann Friedrich Eschscholtz. Seeds were first collected in the Columbia River region and introduced to cultivation by Douglas in 1825.

In the spring of 1824 word came that the Hudson's Bay Company had resolved its organizational difficulties and was prepared to transport, house, and protect a Horticultural Society collector for an indefinite stay on the Columbia River, though warning that he might find "the fare of the country rather coarse and be subject to some privations." The Company's ship, the

William and Ann, was to leave London during the summer and proceed via Cape Horn to the American Northwest.

No time was lost preparing Douglas for the great adventure. Although the area was almost completely unbotanized, Sabine prescribed an intensive study of three texts: Pursh's *Flora Americae Septentrionalis,* which, as we have noted, described most of the plants brought back by the Lewis and Clark expedition; Nuttall's *The Genera of North American Plants,* and Michaux's *The North American Sylva,* the authoritative book on conifers, the predominant tree type of the Columbia River region.

It's possible that John Lindley was assigned to supervise Douglas's studies. Lindley, exactly the same age as Douglas, was an assistant secretary of the Horticultural Society. Originally he had been employed by the council "to have superintendence over the collection of plants, and all other matters in the [Chiswick] Garden," but in these duties he'd long since been shouldered aside by Secretary Sabine. As a consolation, he was occasionally permitted to write descriptions of new plants for the Society's journal when Sabine was busy or couldn't be bothered, and so was unobtrusively beginning to build a botanical reputation.

Born the son of an unsuccessful nurseyman, John Lindley, after a few years of formal schooling, had, like Douglas, been put out to apprentice as a gardener. In 1819, at the age of twenty, he came to London and obtained a clerical job in the library of the great Sir Joseph Banks. Within a year Sir Joseph was dead, but the reflected glory of having worked for him clung to every ex-employee, and Lindley soon found a post as a salaried assistant at the Horticultural Society.

Industry and painstaking scholarship were the qualities most admired in John Lindley, as well as courtesy and gentle manners. But gentle or not, he had a way of bouncing out of treacherous situations and coming out on top. Certainly he outmaneuvered Joseph Sabine when the inevitable showdown came. He continued on with the Society for another thirty-three years, most of them as secretary, to become "possibly the greatest servant it has ever had." He also rose to prominence as a botanist and has been called by a recent authority "the dominant personality in Botany of the early and mid-Victorian era."

If Douglas studied with Lindley that summer, he was in luck. We know that a sincere regard developed between them: later, on the Columbia River, Douglas was to demonstrate more than once his esteem for the rising young botanist. When the predictable break came and Douglas was driven to cut himself off from this friendship too, there is no question but that he was the loser.

In addition to his studies in botany, Douglas was required to familiarize himself with other sciences. Since the Hudson's Bay Company was offering hospitality to only one collector, both zoological and geological enthusiasts had applied for a share of Douglas's time. Accordingly he was instructed in taxidermy, so he could preserve the skins of birds and animals, and also in the structure of rocks, with special emphasis on the identification of fossils.

The time had passed when geologists naïvely believed that segments of shells embedded in mountain sides had been deposited there by Noah's flood. Too many fossils had been found higher than those on Mount Ararat, whose inundation, on biblical authority, had provided the high-water mark of that deluge. Geologists—at least most of them—now believed that Noah's flood had been merely the last, and possibly the mildest, of many such catastrophes that had swamped the world from the beginning of time, wiping out all living things and putting God to the trouble of creating new sets and forms of men, animals, and plants. It was hoped that Douglas, perhaps somewhere in the Rockies, might find evidence of still loftier heights to which the waters had risen during the untold series of cataclysms.

At some stage in the preparations, Joseph Sabine stepped in personally to take over—probably when the time came to visit distinguished travelers. The first call was made on Dr. Archibald Menzies, the pioneer investigator of Pacific Coast flora, now, at the age of seventy, living in retirement in London. A surgeon-botanist on Captain George Vancouver's flagship some thirty years before, Menzies had been the first to see with educated eyes the enormous trees that had captured the imagination of the estate owners of England—the redwoods, the giant oaks and maples, and especially the towering conifer that was one day to be called the Douglas fir. He had also beheld the madroña, toyon, salal, and other plants without precedents in botanical history.

As a ship's doctor, Menzies's excursions ashore had been limited; yet he had collected and pressed enough specimens to awaken the hunger of both the botanists and gardeners of Europe. The Lewis and Clark expedition, visiting some of the same territory a decade later, had whetted appetites even further by bringing back additional glorious plant samples.

Douglas next went calling on Dr. John Richardson, recently returned from the Polar seas and preparing for a second expedition. The British Admiralty, after three hundred years of dreaming about an elusive, ice-free Northwest Passage, had finally decided to explore the frozen wastes above Canada to find out what actually was there. Dr. Richardson was the physician-naturalist connected with one such expedition. Just what he had learned in the Arctic wastes that could be of botanical value to Douglas on the Columbia River isn't quite clear, but Sabine was taking no chances.

Finally there were a succession of visits to botanical and horticultural

PSEUDOTSUGA TAXIFOLIA (*Douglas fir*)

First described by Archibald Menzies (as *Pinus taxifolia*) in 1792. Seeds collected along Columbia River and introduced to cultivation by Douglas in 1825.

specialists: to Aylmer Bourke Lambert, expert on pines, who hoped Douglas could send back enough new species to warrant another edition of his book; * to a Mr. Wells of Redleaf, Kent, compiling information on bee-attracting flowers; to Dr. Friedrich von Fischer, botanist to the Czar of Russia, text writer on the genera of grasses.

No less an authority than the great Linnaeus, father of modern botany, once observed; "Where there is to be found one discoverer, we have one thousand compilers." So now the compilers focused on Douglas, each trying to solicit enthusiasm for his pet project, each seeing in his mind's eye

* A wealthy scientific dilettante, Lambert was determined to go down in history as the ultimate pine authority. His name is unknown today except in *Pinus Lambertiana*, the great sugar pine that Douglas named in his honor.

a different Columbia River region, forests of pines, meadows swarming with bees, or rolling plains of grass, depending on his specialty.

How did Douglas react to all this attention? It would seem safe to assume that, with so many distinguished people investing so much hope in him, his confidence and self-assurance would grow. Therefore it comes as a shock to learn—from the only eyewitness account we possess—that the opposite seems to have been the case.

The eyewitness was Thomas Andrew Knight, president of the Horticultural Society. Understandably delighted by the success of Douglas's fruit-collecting mission to the United States, he wanted to congratulate the young man personally. Sabine tried to discourage a visit to Chiswick, giving as his reason that "Douglas will be terribly frightened," but Knight went anyway. In line with Sabine's prediction, he found the young gardener "the shyest being almost that I ever saw." It wasn't "until I had talked to him for some time in a friendly and familiar way" that Douglas's tensions eased somewhat.

What are we to make of this? Are we to believe that the David Douglas who had talked with increasing confidence to prominent men in the United States was suddenly turned into a tongue-tied clod in the presence of a kindly English gentleman? It seems unlikely. A more plausible explanation is that he put on an act (and acting is strongly suggested in Knight's puzzled description) because Joseph Sabine expected it of him.

Why?

We can only speculate, but perhaps the Secretary, alarmed by signs of self-assertion after Douglas's return from America, strongly suggested that he wouldn't help his career with the Society by acting above his station. Knight may have visited Chiswick shortly after this warning, and as a result Douglas was careful to play the humble retainer, even perhaps exaggerating the role.

We can well imagine that, with the expedition to the Columbia River a reality at last, Douglas was going to let nothing stand in his way. If he'd been required to touch his forelock to Knight like a baronial serf, he undoubtedly would have complied. Sabine's petty tyrannies became less important as the day for the ship's departure grew closer.

On July 25, 1824, Douglas boarded the *William and Ann* in the Thames estuary. He was delighted to find on board Dr. John Scouler, whom he'd known as a medical student at Glasgow, and who had been appointed the ship's surgeon-naturalist through the good offices of their mutual friend Professor Hooker. Douglas was particularly pleased because Scouler immediately accepted him as a scientific colleague and equal.

Despite his democratic attitudes and impeccable medical qualifications, one aspect of the physician might have made a modern traveler, faced with an eight-month voyage under his medical care, understandably nervous. Dr. Scouler was just nineteen years old.

On December 16, after the usual stormy passage around Cape Horn, the *William and Ann* lay off Mas-a-Tierra, the inner island of the Juan Fernández group. Douglas and Scouler looked out at the rocky coast with interest, reminding each other that this was where Robinson Crusoe was washed ashore after the famous shipwreck.

In actual fact, Alexander Selkirk, the prototype of Robinson Crusoe, wasn't shipwrecked and was never washed ashore. He was a cantankerous Scotsman, a ship's mate, who was marooned by his equally hot-tempered skipper after an argument. The year was 1704, but even then vessels habitually called at the island for water and wild fruits. Neither Selkirk nor his captain thought the mate would have to wait more than a few weeks for a ship. As it turned out, owing to wars, storms, and sheer bad luck, it was four years and four months before rescue came.

Selkirk had lived so comfortably on the well-stocked island that when he returned to England it never occurred to him that anything unusual had happened—not, that is, until a publisher asked for his recollections at a penny a page. At those rates, Selkirk found that he'd undergone many harrowing, page-filling experiences. The resultant pamphlet was far from a best seller, but it served its historic purpose when it fell into the hands of Daniel Defoe.

Douglas and Scouler went ashore and were delighted to find a romantic young Englishman camped on the island. Bent on reenacting Crusoe's experience, he had jumped ship at Valparaiso and reached Mas-a-Tierra with some Chilean cattle hunters. Douglas never saw the hunters, though he heard their shots. The men were across the hills slaughtering wild animals said to be the descendants of bulls and cows left by Captain George Vancouver.*

Scouler and Douglas turned their attention to botanizing. They worked through the lush vegetation for two days, collecting, drying, and pressing specimens of seventy different kinds of palms, ferns, and flowers. Douglas made no attempt to collect roots; there was nothing growing in that year-round balmy climate that would last outdoors through a single English winter.

In one shady ravine he found a stand of low-growing ferns too tempting to resist. "I laid myself down on a carpet of it close by a crystal rill descending through the rugged but beautiful hills," he notes in his journal. It

* Vancouver seems to have made a practice of planting cattle (obtained in California) on Pacific islands to provision ships that came after him. Spaniards generally left goats; for instance, on Goat Island in San Francisico Bay.

is no wonder that he hated to say good-bye to "this classic island—the Madeira of the South."

His reaction to the Galápagos Islands, the ship's next point of call, was very different. The black, rocky ledges rising abruptly from the sea, with but "little herbage," were bereft of grace or charm. He found the contour of James Island, as the *William and Ann* moved in to anchor, harsh and inhospitable—"mountainous . . . rugged, with . . . vestiges of volcanic craters and vitrified lava."

The islands didn't improve on closer acquaintance. Douglas and Scouler went ashore, a shore described by a subsequent observer as "of hideous black lava that had been twisted and buckled and tossed about as though it were a petrified stormy sea." They found lizards and tortoises everywhere, creeping or crawling or lying stupefied in the sun. Pulling himself together, scientifically and gastronomically, Douglas noted a tortoise "weighing 400 lbs; a lizard, 3 feet long, of a bright orange-yellow; both good eating."

Presumably he determined the weight of the tortoise when it was swung aboard ship. For three days the crew of the *William and Ann* captured the enormous creatures, storing them on deck, belly up, for future use.

Strangely, Douglas seemed at first reluctant to botanize in the midst of some of the most bizarre flora the world has to offer. He seemed disoriented by its very difference, upset that it fit into no familiar pattern. "Some of the trees in the valleys are large," he notes, "but . . . few of them were known to me." He also saw some cacti and other prickly plants which he could not identify. They didn't appeal to him, but he concedes that a professional botanist might find "many of them, no doubt, interesting."

Finally realizing that he was in an area where "everything, indeed the most trifling particle, becomes of interest in England," he methodically worked through the peculiar vegetation, collecting botanical specimens.

Dampness plagued him; in that humid atmosphere astride the equator his specimens refused to dry. Tropical downpours alternated with periods when "the sun broke through and raised a steam from the ground almost suffocating." All but forty of some one hundred and seventy-five specimens were spoiled by mildew; of those he salvaged, he could only comment, "What they may be I cannot say."

Scouler too, trying to acquire a collection of birds and lizards, was harried by the weather. Specimens were easy to obtain—the birds "so little acquainted with man's devices that they were readily killed with a stick"—but preserving them was another matter. Of the forty-five skins put out to dry, all but one rotted in the saturated atmosphere of James Island and in the humid air that continued to follow the ship as it progressed north.

"Never in my life was I so mortified," Douglas declares. He is referring to the loss of his specimens but, to modern ears, he might seem to be bemoaning a more significant lost opportunity.

Of course it is easy for us, with hindsight, to smile at those early nineteenth-century naturalists who were blind to the abounding evolutionary clues—clues that so forcibly impressed themselves on the young Charles Darwin when he visited the Galápagos some ten years later. But neither Douglas nor Scouler was prepared in any way for evolutionary ideas. They considered themselves men of science, it is true—investigators of objective realities; but first they were children of God, and His truths must always prevail over fallible human perceptions.

In Genesis the Creator had clearly stated, "Let the earth bring forth grass, the herb yielding seed, and the fruit tree yielding fruit after his kind, whose seed is in itself." Botanically, this was taken to mean that plants were fixed and unchangeable in nature. A rose was a rose and a strawberry was a strawberry: they always had been and always would be. Horticulturally, man had improved both the rose and the strawberry by grafting and by selection of seedlings—biblically permitted because God wanted man to make maximum use of His gifts—but each plant was still clearly "after his kind, whose seed is in itself."

Even to suggest what modern botanists since Darwin have taken for granted, that roses and strawberries belong to the same plant family—meaning that, millenniums ago, they shared a common ancestor—would have seemed to Douglas and Scouler botanically shocking; almost as shocking, in fact, as the zoological perception which outraged the Victorian world, that man shared a common ancestor with apes.

We have noted that one of Thomas Nuttall's botanizing objectives was "to admire the wisdom and beauty of creation." The phrase might well serve as a text to cover much pre-Darwin scientific investigation. Douglas and his contemporaries believed that their purpose was to reveal the perfection of God's completed world, not to suggest that the scaffolding was still up and His work still in progress. If this approach sometimes led to puzzling anomalies—such as random boulders from Scotland on the chalk downs of Kent or hippopotami bones in Siberia—it was only because man's mind was unable to penetrate to the full light of divine wisdom. Douglas desperately clung to this mystic position when, years later, finding the foundations of his scientific world beginning to shake, he declared and reiterated his determination never to yield to the "philosophical scepticism which makes us deny the reality of what we have not seen and doubt the truth of what we do not perfectly comprehend."

Of course Charles Darwin, too, began as a disciple of the same scientific age. But he was better prepared to accept the astonishing perceptions of his intellect on the Galápagos Islands in 1835. His grandfather had been an evolutionist—true, on the somewhat romantic plane popular at the end of the eighteenth century—but at least Darwin had some foundation in the concepts of change and progression. Even so, the implications of his theory were so disturbing that years passed before he was able to commit himself to publication. Fear haunted him, fear of the public storm, "the howl of execration" he knew would break when he shattered the myth of divine creation. For decades he postponed the inevitable, busying himself with minor projects, taking refuge in invalidism. It wasn't until 1859, more than twenty years after his voyage to the "crucible of evolution," that *The Origin of Species* was published, the book that was to change man's concept of his world and of himself.

For eight months, the officers, crew, and passengers of the *William and Ann* had remained disgustingly healthy, but finally, off the Mendocino coast of California, Dr. Scouler got his chance. "Furious hurricanes," Douglas noted in his journal, "a thousand times worse than Cape Horn." At the height of the storm, the second mate fell from the rigging and fractured his right thigh.

He was carried below, and Scouler went to work. We suspect that this was the young doctor's first exercise in orthopedic surgery, since he describes his treatment in detail, noting in *his* journal that he adopted "Pott's plan in preference to that of Desoult." The operation was made difficult by the buffeting of the ship but, on the whole, Scouler was satisfied that he had acquitted himself well. Douglas was less concerned with medical technique when he noted about the patient, "The excruciating pain which this poor man suffered until the termination of our voyage can hardly be expressed."

Finally the storm abated and the ship ventured in toward the coast. On April 2, 1825, the woody promontory known as Cape Disappointment was sighted at a distance of about thirty miles. Several more days were consumed tacking off shore in adverse winds, but on April 7 the *William and Ann* edged in toward a collection of shoals and sand bars that gave no indication of a river anywhere in the vicinity. "The current is very rapid," Douglas wrote, "and when the wind blows from the west, produces a great agitation. The water on the sandbar breaks from one side to the other so that no channel can be perceived."

He was describing the protective camouflage by which the fabled River of the West had concealed itself from white men for almost three centuries.

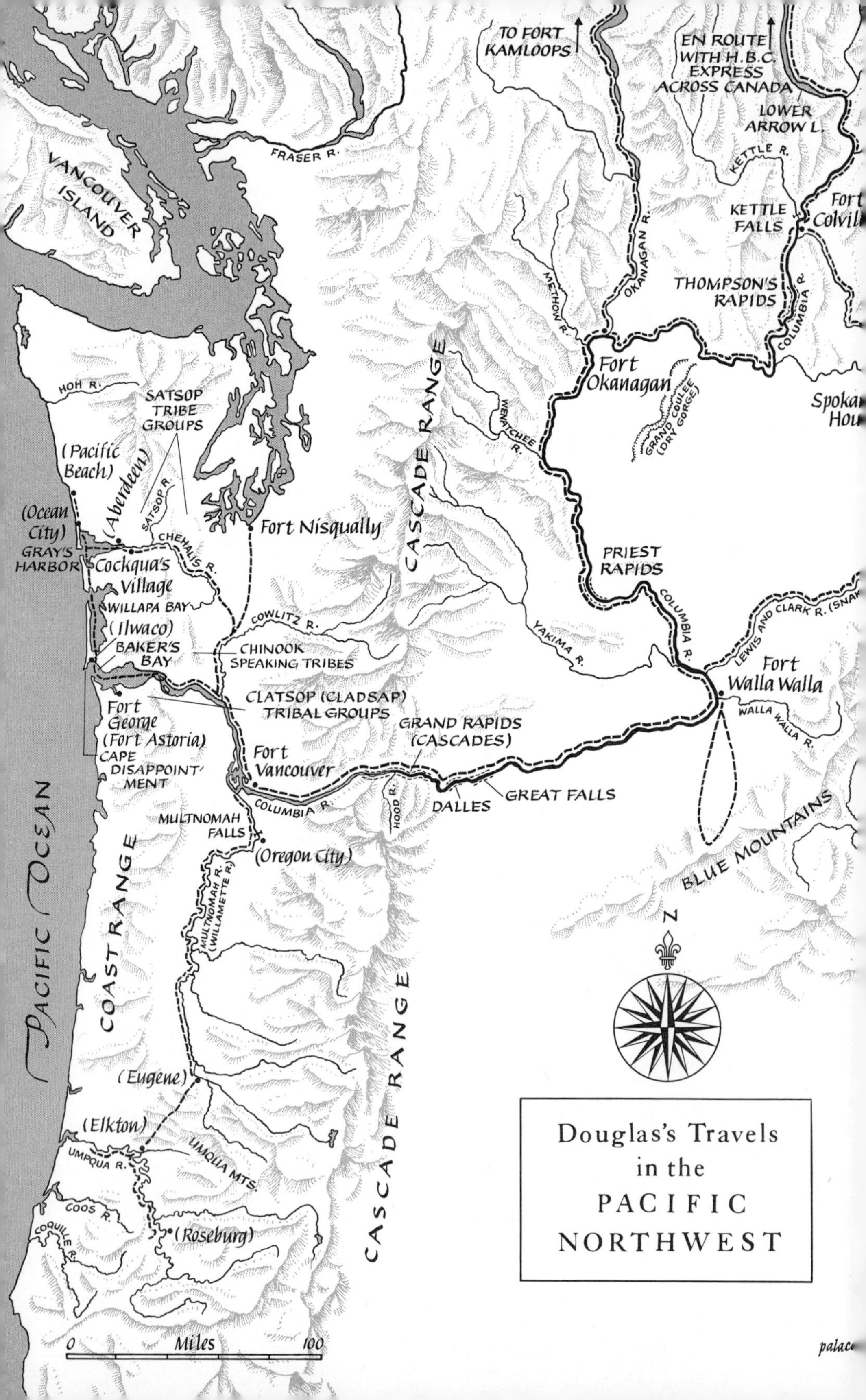

Douglas's Travels in the PACIFIC NORTHWEST

4

THE GREAT RIVER

MEASURED IN TERMS OF VOLUME OF water delivered to the ocean, the Columbia is the third largest of North American ice-free rivers. An average of 280,000 cubic feet per second issue from the Columbia, in comparison with 400,000 from the Saint Lawrence, and 513,000 from the Mississippi.

But "average" flow of water hardly does justice to the Columbia's potential. In early summer, when the snows melt in the Rockies and the tributaries are gorged, the Great River, running in places three hundred feet deep and driven with the power of fifty million horses, can push a column of fresh water four miles out to sea. At such times, the swollen capacity of the Columbia climbs to 1,200,000 cubic feet per second, a rate of delivery that places it, if only temporarily, fourth among the rivers of the world, exceeded in volume only by the Congo, La Plata, and the Amazon.

It seems incredible that so large a river should have gone for so long undiscovered. Two hundred and seventy-two years were to elapse between Magellan's triumphant burst into the Pacific and the first nervous entrance of a sailing ship into the Columbia.

Admittedly, the early explorers along the Pacific coast weren't looking for a river. They were after bigger game—the western terminus to the Northwest Passage, that mythical sea lane supposed to connect the Atlantic to the Pacific. Francis Drake (not yet knighted) was instructed to search for it in 1578, but Drake always preferred looting Spanish galleons to exploring, and it is doubtful if he sailed much farther north than the bay in California which bears his name. Certainly the Spanish navigators of the period—Cabrillo, Aguilar, Vizcaíno—found no passage for all their scurvy-harassed expeditions.

Traffic off the western coast subsided for more than a century. When it resumed—as it did with a rush in the 1770s—it was fortune, not fame, that lured the ships back. For some time it had been suspected that the Russians were up to something both mysterious and lucrative off the coast of Alaska. Captain James Cook of the British navy, sent to investigate, stumbled on the secret commerce more or less by accident. His sailors, trading with Indians up and down the coast to pass the time, swapped brass buttons for sea-otter skins. Later, in Canton, the jolly tars found that they could sell the skins for fabulous prices to the Chinese mandarins, and the rush was on.

During the peak of the sea-otter boom—the decade from 1780 to 1790—ships from Spain, England, France, Russia, and the fledgling United States congregated by the dozen in the chilly waters off northwest America. The genial and trusting sea otters were clubbed to death by the thousands, their furs shipped to Canton to be traded for teas, silks, spices, and porcelains—luxury goods commanding top prices in the various home markets. Profits could be enormous; it was said that a Boston skipper-owner could retire for life after a single successful tour of the circuit, peddling trinkets for furs for teas and silks.

Oddly enough, in the midst of this hectic commerce, there was a revival of interest in the Northwest Passage, but suppositions as to its nature had changed. The waterway was now supposed to be a great river—the River of the West—which flowed from the Great Lakes through the Rockies and came out at the Pacific Ocean. But where?

That question seemed answered in 1775 when Bruno Heceta, an exploring Spaniard, felt the current of an enormous river running under his ship. Owing to shoals and breaking water he could not actually see the estuary, and he wasn't willing to risk his ship among the sandbars, especially in the inclement weather. So he named the unseen river the San Roque, charted the latitude of its mouth as N. 46° 9′, and sailed away, leaving the honor of actual discovery to the next explorer.

The next explorer was a voluble busybody of a ship's captain named John Meares. In 1788, Meares arrived on the coast and, Heceta's chart in hand, examined the location at which the River of the West was supposed to debouch. He was unable to get close—bad weather again—but that did not prevent him from reaching conclusions that he published on his return to England. "We can now with safety assert that no such river as the Saint

Roc exists," he declared. To prevent future explorers from wasting their time, he named the churning waters at the sandbars Deception Bay and a prominent headland Cape Disappointment.

That was how the situation stood when Captain George Vancouver arrived on the scene in command of two ships of discovery in the spring of 1792. He had studied Heceta's chart and read Meares's book, and neither made him hopeful of finding an important estuary at north latitude 46° 9′. Still, he owed it to the British Admiralty to have a look before proceeding north to what, in his opinion, were more likely locations.

George Vancouver was an experienced navigator and nobody's fool. He lay off Cape Disappointment for some time studying the discolored water that sifted with force through the sandbars. He tasted it; it was fresh. But a possibility occurred to him. Suppose the flow was not from a single large river, but from a series of smaller streams? That would dispose of the theory of the great inland waterway, while still accounting for the strong current.

There was, of course, only one way to find out for certain—sail in and have a look. But the weather was bad again and Vancouver was afraid to risk his ships. He decided to come back in the autumn when the winds tended to subside—that is, if he hadn't found the River of the West in the meantime.

He sailed north and around the middle of April met up with the American merchant vessel *Columbia*, commanded by Robert Gray. Vancouver was already acquainted with Captain Gray; he knew him to be a resourceful and courageous seaman noted, among other things, for having first carried the Stars and Stripes around the world. As the ships dipped flags and sailed on, Vancouver may have had misgivings when he saw the direction in which Gray was headed—toward the coast he had just left. But the misgivings couldn't have lasted. Perhaps a hundred ships had searched up and down that shore in the past decade. They hadn't found anything and neither would Gray; no merchant captain would be foolish enough to risk his ship among those sandbars.

Vancouver sailed northwest, entering Juan de Fuca Strait and moving up around east of it. All summer he crept past islands and quartered around capes, searching constantly for river outlets. He found none big enough to qualify as the River of the West but, by September, he had sailed north into the Pacific again, thus discovering by circumnavigation the existence of a large island—the present Vancouver Island, British Columbia.

Well pleased with his summer's work, Vancouver moved his ships down to Nootka harbor, a provisioning port which also served as a rendezvous and gossip center for trading vessels. He saw the *Columbia* at anchor and, in course of time, Captain Gray rowed over to exchange news. Quite casually, considering it more an interesting experience than a historic event, Gray passed along the information Vancouver had been dreading to hear.

He had entered the great river. It had been a difficult approach—the weather was bad—but he finally pushed in past the reefs. Finding a channel, he threaded through the sandbars and sailed into an astonishing estuary, five miles wide at its mouth. He sailed some distance upstream, then anchored to trade with the Indians who swarmed out of the forest at the novelty of seeing a white man's ship on their river. Gray added some one hundred and fifty sea-otter pelts and three hundred beaver skins to the furs already in his hold before edging out into the ocean again. He told Vancouver one more thing: he had named the river the "Columbia" after his ship.

Dismayed, Vancouver hurried back to the scene of the discolored water, carrying a chart Gray had sketched for him. Still unwilling to risk his flagship in those breakers, he sent in his smaller vessel, the *Chatham*, Lieutenant Broughton commanding, to explore. Broughton found the channel and maneuvered through the sandbars to reach the majestic, silent flowing river. He sailed up it a hundred miles, where Captain Gray had sailed possibly twenty-five. He went ashore at the terminus of his voyage, planted the Union Jack and took possession of all surrounding lands in the name of His Majesty, King George III of England. He moved slowly back down the river, charting islands, promontories, inlets, and tributaries. In his journal he described Indians, vegetation, and animals.

In short, he did all the correct and legal things to establish British rights and claims to discovery. But in the end it was wasted effort. Robert Gray had done few of these things; but on the murky morning of May 11, 1792, he had been the first to sail into the River of the West. When he left it, it was the Columbia—a myth and a mystery no longer, ready now for orderly exploration and a new destiny.

The *William and Ann* entered the river late in the afternoon of April 7, 1825, and dropped anchor in Bakers Bay, a sickle-shaped body of water in the lee of Cape Disappointment. The Hudson's Bay Company post in the area was still some distance off but, since dusk was approaching, the captain settled for firing a cannon to announce his arrival, and then ordered extra rations of rum for all hands. After eight months and fourteen days of "tedious voyage" the celebration was welcome. "We spent the evening with great mirth," Douglas reports, "and at an early hour went to sleep, to sleep without noise and motion, the disagreeable attendants of a sea voyage."

Next morning, when no one appeared from the post and heavy rains and fog prevented the ship from proceeding upriver, Douglas and Scouler, impatient to range "through the long wished-for spot," borrowed a boat and rowed ashore. The first hours of botanizing in those virgin woods were to

GAULTHERIA SHALLON (*salal, lemon-leaves*)

Originally collected by Archibald Menzies in 1792, Douglas describes his delight at recognizing the shrub on the Columbia River in 1825 in the following words: "On stepping on the shore *Gaultheria Shallon* was the first plant I took in my hands. So pleased was I that I could scarcely see anything but it."

be indelibly impressed on Douglas's mind. "Gaultheria shallon (salal) was the first plant I took in my hands," he recounts. "So pleased was I that I could scarcely see anything but it. Mr. Menzies correctly observes that it grows under thick pine-forests in great luxuriance and would make a valuable addition to our gardens."

Roaming farther, he found blueberry bushes in bloom and, close to the ground, clumps of coral bells and saxifrage. But dominating his consciousness were the enormous trees that loomed above and around; some spruce and hemlock, but predominantly Douglas firs. "The trees which are interspersed in groups or standing solitary in dry upland . . . are thickly clad to the very ground with widespreading pendent branches, and from the gigantic size which they attain . . . form one of the most striking and truly graceful objects in Nature," he tells us. "Those on the other hand which are in the dense gloomy forests . . . the trunks being destitute of branches to the height of 100 to 140 feet . . . arrive at a magnitude exceeded by few if any trees in the world." He gives the dimensions of one fallen giant as "entire length 227 feet, 48 feet in circumference [more than 15 feet in diameter] 3 feet above the ground."

It is not possible for us to share now the awe and wonder of those two young naturalists at their first encounter with that primeval forest. Even by entering a "dense gloomy" stand of timber today we can get only inklings of what they must have seen and felt. We can still move in the half light under a high canopy of branches, listening to the twittering of birds come down to us from the "roof" above. We may think that in the eerie shadows, with the duff crackling beneath our feet, we have successfully recaptured the past, until we are startled by huge stumps–five, seven, even ten feet in diameter–left by loggers perhaps a century ago. Suddenly we realize that the quite sizable trees we see about us are saplings compared to the giants Douglas and Scouler saw that April day in 1825.

They returned to the ship to find Indians aboard. The tribes of the lower Columbia were by this time accomplished hucksters, experts at catering to the white man's needs. They knew, for instance, that after a sea voyage fresh foods were in demand. Douglas notes that they were offering "salmon, fresh sturgeon, game, and some prepared roots with dry berries," and that the Indian salesmen "soon showed themselves to be dexterous people at bargaining."

Douglas also imagined that "the natives viewed us with curiosity." Possibly so, but it couldn't have approached the curiosity with which he and Scouler viewed the startling Chinook physiognomy for the first time. Douglas, in fact, is a bit incoherent in his description: "The practice of compressing the forehead, of perforating the septum of the nose and ears with shells, bits of copper, beads, or in fact any hardware, gives a stranger a curious idea of their singular habits."

The practice of compressing the forehead, which resulted in a sloping skull, always excited debates among whites. The Chinooks prided themselves on the backswept look, accomplished by passing their first year of infancy between confining boards. They considered it not only aesthetically pleasing but a sign of caste distinction; slaves, for instance, were not allowed to shape their children's heads in this manner.

On the whole, as whites got used to the look, they endorsed it as adding dignity and nobility to Chinook men. Chinook women had other assets which needed no endorsement. From the time of Captain Gray's sailors, white men had found the girls of the lower Columbia irresistible.

When the rain and fog finally lifted, the *William and Ann* moved cautiously upriver. No word had yet been received from the trading post called Fort George, and, though the possibility of massacre seemed remote, all on board were understandably nervous.

Douglas went ashore with the landing party. Climbing the hill to the stockaded building—at the site of the present city of Astoria, Oregon—they discovered the reason for the ominous silence of the past few days. The fort was being abandoned in favor of a new trading post; most of the garrison had already moved, leaving behind a skeleton force unable, for want of a cannon, to answer the ship's salute.

The Hudson's Bay Company's new headquarters were under construction a hundred miles upstream, at approximately the site where Lieutenant Broughton had planted the Union Jack thirty-three years before. With unrelenting loyalty to the man who was so close and yet so far, the new establishment was being called Fort Vancouver.

The men of the Hudson's Bay Company, a corporation chartered in 1670 by Charles II so that his "dear cousin, Prince Rupert" could do a bit of fur hunting in an area stretching out over about three-quarters of the present Dominion of Canada, were old hands at building trading-post forts. In the course of a hundred and fifty years, they had built them on prairies and tundras and deep into the Rockies. Now, with new territories, recently acquired, that extended all the way to the Pacific, they were hard at work developing forts up and down the Columbia River.

Construction was well advanced when Douglas reached Fort Vancouver late in April. A palisade some eighteen feet high enclosed an area of approximately two acres. Living quarters for a hundred and fifty people had been built and occupied; bakeries and kitchens were in operation. The workshops of carpenters, smiths, coopers, wheelwrights, millwrights, tinsmiths, and other artisans were putting out goods and services. Warehouses stored supplies and furs accumulating for shipment; and of course there was Indian Hall, where beaver skins might be traded for beads, knives, hatchets, and even such luxury items as flintlock muskets and three-cornered hats.

Douglas was welcomed at Fort Vancouver as an honored scientific guest. "Every attention and assistance" was offered him and though he was lodged in a tent for awhile, owing to a housing shortage, he finally graduated to quarters especially built for him—"a hut made of bark of *Thuya occidentalis* [Arborvitae]"—with which he was delighted.

Another thing that must have delighted him, and made him feel thoroughly at home, was the Caledonian atmosphere. Almost to a man, the officers at Fort Vancouver were either Scots or of Scottish descent.

It has been said about the English that the Irish fought their wars and the Scots built their empire. Certainly more than 75 percent of the Hudson's Bay Company field personnel spoke with the burr of the tartan. Poverty originally drew adventurous lads from the Highlands and Orkneys to the wilds of Hudson's Bay, but, by the time David Douglas knew the Company, service in it was a proud tradition. Few recruits now came directly from the Land o' Cakes; the new generations of apprentices tended to be the sons and grandsons of employees, now Canadian in nationality but still strongly Scottish in traditions. Kilted Highlanders stood guard duty outside the principal Hudson's Bay Company forts, and bagpipers paraded for distinguished guests in the mess halls.

There was a hierarchy, with its etiquette, at Fort Vancouver, as at all the major establishments. First in authority was the chief factor, holder of a full share in the annual distribution of dividends from London, in charge of a territorial division of the Company's empire. Next in rank was a chief trader, owner of a half share, usually presiding as officer in charge of a trading post. Of lesser rank, but still counted as gentlemen, were the salaried clerks—ranging from seniors, earning as much as £150 ($3,750) while they waited for an opening as chief trader, to lowly apprentices serving out their terms at £20 ($500) per annum.

At the next level, eating at a separate table in the mess hall, were the middle employees, presided over by the postmaster. (His title stemmed from supervising work at the post, not mail.) At this table sat the artisans, interpreters, and French-Canadian boatmen, the fabled *voyageurs*.

These "troubadours of the lakes and rivers" were tireless oarsmen who could paddle canoes or row boats for as many as eighteen hours a day, pausing only for a five-minute pipe every two hours and rations of cold pemmican at noon. They rarely aged enough to retire, usually dying of heart failure or pneumonia before the age of fifty. They were known for their simplicity, gaiety, and willingness to perform phenomenal work, for which they were paid £17 ($425) per annum.

Finally, there was the laborers' table which, at Fort Vancouver, was populated by Iroquois imported from across the Rockies and Kanakas from the Sandwich Islands, since the tribes of the lower Columbia never took kindly to physical work.

The same food was served at all three tables, though grades of tea were different. At the conclusion of a meal, at a signal from the postmaster, the lower orders rose and left the room while the gentlemen lingered on to enjoy wine and tobacco. We may assume that David Douglas, relaxing in such company, first lost his grip on "singular abstemiousness" at the Fort Vancouver mess.

Presiding officer and divisional head of the entire Columbia Department was Chief Factor Dr. John McLoughlin. He was an impressive figure, standing six feet four inches, with deep-set gray eyes under a cascade of prematurely white hair which he parted in the middle. Though only forty, he cultivated a patriarchal manner which he fortified by always wearing black clothes and carrying a gold-headed cane. When Douglas knew him, he was about to begin the second phase of a remarkable career that was to span the decline and fall of the Hudson's Bay domination of the Columbia River and terminate in his own uneasy Americanization.

But in the spring of 1825 Dr. McLoughlin had more immediate concerns than the future course of empire; he was considering ways and means of turning a profit at Fort Vancouver. The previous year—though under different management—the entire Columbia Department, from the Rockies to the ocean, had yielded fewer than twenty thousand beaver skins, a poor showing in comparison with production in the east.

The problem, as the doctor analyzed it, was that the Indians refused to bring in furs because they'd lost respect for the white man. The greed and treachery of the Canadian traders who had preceded the Hudson's Bay Company into the area had thoroughly disgusted them. The infamous "smallpox bottle," used to blackmail tribes into stepping up fur production on threat of uncorking the dreaded disease, was only one example of unprincipled behavior.

In order to win back the Indians' respect, McLoughlin instituted a new rule of law, just and firm, applicable to red and white man alike. When the tribes realized that he meant it, when they saw Company employees disciplined for offenses against them, they were completely won over. A new era of confidence began, with Indians once again trusting ordinary white men and revering Dr. McLoughlin. They named him the White-headed Eagle, carved him into their totem poles, named children after him, and even stopped killing slaves on ceremonial occasions because the doctor didn't like it. But the one thing he most wanted they didn't succeed in giving him—increased fur production. McLoughlin could never understand it.

Douglas, fitting into the routine of Fort Vancouver, began making botanizing trips into the surrounding country. Scouler, who had come up-river while the *William and Ann* unloaded, frequently joined him. Together they searched the forests, the glades and meadows, and the sandbanks along the river.

There is something infinitely appealing about the picture—the two young men in the prime of youth, moving lightly in spring days through verdant growth breathtaking in its beauty. From our viewpoint they were too careless of the wonders they saw, too confident that the scene would last forever to note down in exact detail what that pristine landscape looked like. We can only pick impressions from hints and clues in their journals.

We know, for instance, that plant identification was often difficult. Too much was new, or looked new, in that rich vegetation. They could only collect specimens, fold them in paper—"lay them in," as the phrase went—and hope that their botanical books back at the fort would provide clarification. When the texts were ambiguous, debates ensued. What was the blue flowering plant? A snapdragon? It looked something like one, yet missed on several points of identification. Could it be a species of that new genus *Collinsia?* Impossible to tell without seed, and seeds were months away; though Douglas might enter in his journal a cryptic reminder, "Get seed of this."

Sometimes the books failed completely and the men were left with no idea of what they had collected. The repeated dilemma is illustrated in the following extract from Douglas's journal:

(28)* *Ranunculus* sp., annual; flowers small; yellow.

(29) ——(?) perennial; sandy soils; abundant.

(30) *Trientalis americana*, in shady pine woods among moss; abundant.

(31) *Smilacina racemosa* (?), perennial; shady woods, in rich vegetable soil; white flowers; plentiful.

* The digits in parentheses are the numbers assigned exclusively to each plant. All specimen sheets, seed containers, bottled bulbs, and roots, everything connected with the plant, carried the same number to prevent later confusion.

(32) *Smilacina* sp., perennial; a small plant 6 inches to a foot high in the same situations as the former. S[hade].

(33) *Cynoglossum* sp., perennial; flowers fine blue; a strong plant 2 to 3 feet high; in thick shady woods.

(34) ——(?); shady woods among moss; flowers white; 4 to 6 inches high; annual. Can this belong to *Trientalis;* April.

Actually the professional botanists in England preferred that blanks be left when identification was in doubt. They hated wild guessing, claiming that it disturbed the Olympian clarity of mind necessary for their elucidations, but Douglas was never impressed by this argument. He believed, rightly or wrongly, that botanists preferred unidentified specimens so they could name them in honor of their own friends, often other botanists who would be expected to reciprocate in kind. So he continued to bestow names on anything that looked interesting, relying on the rules of nomenclature (which held that a valid name, once given, could not be changed) to protect him from scheming botanists.

The young men seem especially to have enjoyed their last day of botanizing together, just before Scouler shipped out from Fort Vancouver. They had rowed over to an island in the river and were delighted by the unexpected novelty of the flora. Almost immediately they spotted a forget-me-not and, in joyful agreement, named it *Myosotis Hookeri* "after Dr. Hooker of Glasgow." Next Douglas found a distinctive phlox, which he dedicated with the respectful wording, "This exceedingly beautiful species I name *Phlox Sabinii*, in honor of Jos. Sabine, Esq." He named an evening primrose a little less formally *Oenothera Lindleyana*, "after Mr. John Lindley, the secretary." Bowing to present company, he named a monkey flower *Mimulus Scouleri*, "after John Scouler, who has been the agreeable companion of my long voyage from England and walks on the solitary Columbia." No doubt Scouler found some plant in that abundant flora with which to honor *his* companion, but unfortunately both journals are silent on the point.

It is sad to relate that these names fared badly in botanical records. Partly this was Douglas's fault; he was often too quick to pronounce as "new" species already known and described. But even more it was due to the botanical upheavals that followed Darwin. Though Douglas's name was usually preserved as taxonomic system replaced system, the same can't be said for his commemorative candidates. Of all the names gifted out with such enthusiasm that day in May, 1825, not one can now be found in an up-to-date Flora.

In June, Douglas went upriver with one of the brigades that periodically supplied the trading posts in the interior. The spring floods were in full torrent, and as the boats made their way up the famous Columbia River gorges, progress was slow. Douglas was able to keep pace as he scrambled along the shore, botanizing. Above him rose what has been described as "towering rock precipices, the delicate lace of waterfalls tumbling a sheer hundred feet or more, towers, castles, and pinnacles in the colored rock."

Some forty-five miles above Fort Vancouver, the smooth, deep-running river changed character. Broken water commenced as the Columbia, dropping sixty feet in two miles, made its escape from the mountains. William Clark, with the Corps of Discovery, had called the stretch of roiling current the Grand Rapids, a name still used in Douglas's time though the term "Cascades" was beginning to be heard.

Above the rapids—portage around them consumed most of a day—the river settled down again to its smooth and powerful flow for some fifty miles. Then the roar of waters was heard again as the great cataracts neared. First came the Dalles, described by Douglas as a torn and jagged flow of current "through several narrow channels formed by high, barren and extremely rugged rocks." The pounding water, lashed white, foaming, spitting, drove forward with a fury and power that William Clark, watching in awe some twenty years before, could only call "tremendious."

Above the Dalles the river put on its climactic display. The enormous crest narrowed from half a mile in width to less than four hundred yards, then, "swelling, boiling, and whorling," again in the words of Clark, plunged over a cliff in the staggering spectacle known as the Great Falls.

They're all gone now: Grand Rapids, Dalles, Great Falls, names built into the romance and history of an age are gone, buried under concrete dams. Where fierce currents once raged, gauzelike veils of misting water now slide down cement slopes. The river's thundering transit of the Cascade Mountains, a feat almost as remarkable as the mythical transit of the Rockies by the River of the West, has been reduced to a series of orderly lakes, descending in a succession of tranquil steps from one level to the next.

We wonder what David Douglas would think if he could see the roaring waters of his memory now seeping placidly to sea. Superficially he might find it politic to approve, since Genesis is specific in its instructions to mankind to subdue the earth. But on a deeper level he could only be dismayed that his once mighty river had been not only subdued but taught to play dead.

Douglas left the boat brigade above the Great Falls. For a score of miles the river had been flowing through "barren hills, with the greater part of the herbage scorched and dead by the intense heat." It was the lee or rainless side of the Cascades, a country in which he saw little prospect of productive botanizing.

Accompanied by his escorts, a *voyageur* and two Indians supplied by Dr. McLoughlin, he returned to the Grand Rapids where a salmon run was attracting crowds of fishermen, both local and those who had "come several hundred miles to their favorite fishing grounds." He spent hours observing the various netting and trapping techniques employed. The more daring tribesmen worked from planks suspended over the water. They dipped with long poles equipped with nets and periodically came up with a struggling salmon.

"The fish are of good quality," Douglas reported, "much about the same size (15 to 25 lbs) as those caught in the rivers of Europe," a reminder of a bounty that had once been widespread. In Ireland, laborers refused to eat fish more than twice a day during salmon season and, in Scotland, thrifty farmers pitchforked them out of the rivers by the ton to rot in and fertilize the fields.

On the Columbia, salmon season was continuous from May to December. Chinooks and bluebacks ran to mid-August; after that came silver salmon, considered inferior in quality but still very edible. In February a small, smeltlike fish was to be had in quantities and, in March, sturgeons weighing up to a thousand pounds could be caught. When not eating fresh fish, the tribes of the river subsisted on smoked and dried fish. They even employed a kind of marinating process for preserving fresh fish for long periods.

Douglas gives us no estimate of the salmon caught annually at the Grand Rapids beyond the general statement that it was "an almost incredible number," but a contemporary observer puts the catch at one thousand tons–which would work out at about one hundred thousand average-sized (twenty pounds) fish. Later, on the Little Spokane River, a Columbia tributary, Douglas calculated the daily catch at two thousand salmon.

There were many fishing meccas throughout the Columbia system in those days. One of the most popular was at Kettle Falls, near the present United States–Canadian border, where tribes from as far away as Idaho and Montana congregated to catch a year's supply of fish. Salmon swarmed up the Snake, the Yakima, the Okanogan, even penetrating to Columbia Lake, the source of the great river 1,210 miles from the sea. There is no way now of guessing the total number in Douglas's time, but clearly it was in the many millions.

Today an accurate count at Bonneville Dam, through windows on a fish ladder, reveals that the salmon migrating upriver to spawn total less than six hundred thousand. As denizens of wild and free waters, they don't take kindly to escalators.

In July, Douglas journeyed downstream to the river's mouth. On either bank stretched such a landscape as to impress even the most insensitive observer. An officer aboard the *Columbia* during the first epic days of discovery had noted, "the land was beautyfully divercified with forists and green verdant launs." Farther upstream, low hills rolled back from the river, covered with "clumps and copses of pine, maple, alder, birch, poplar, and several other trees." Thirty-three years later, Douglas adds another tribute: "The scenery in many places is exceedingly grand . . . the most part covered with wood, chiefly pine . . . with a thick herbage of herbaceous plants."

The scenic enjoyment didn't end with the vegetation. Seals and sea otters played around islands in the river, and, as Douglas's canoe descended, flocks of wild ducks, geese, and cranes rose from the water, circled, fluttered, and settled down again.

Douglas's main purpose in coming downriver was to search for a *Cyperus* tuber which the Lewis and Clark journals claimed to be the potato of the Columbia tribes. This information surprised Douglas since, by his own observation, a bulb called "quamash"—meaning "sweet" in Chinook dialect—seemed to supply the Indians' need for a starchy vegetable. He never found the *Cyperus* tuber, but continued to eat quamash throughout his travels. "Taste[s] much like a baked pear," he records. His only objection to the universal vegetable was aesthetic, since quamash tended to produce flatulence. Once in an Indian lodge he "was almost blown out by strength of wind."

Douglas was becoming increasingly interested in the Indian tribes. By midsummer he could speak enough Chinook to make himself understood, and was able to distinguish between other clans of the lower river. "Most of the tribes on the coast—the Chenooks [Chinooks], Cladsaps [Clatsops], Clikitats [Klickitats] and Killimucks [Tillamooks]—from the association they have had with Europeans are anxious to imitate them and are on the whole not unfriendly. Some of them are by no means deficient of ability. Some will converse in English tolerably well, make articles after the European models, [though] they are much prejudiced in favor of their own way of living."

He visited Indian villages—communities of four or five, though sometimes as many as eight, lodges. Lodges varied in size, but a building sixty feet by twenty feet, constructed of split planks, seemed about average. Three

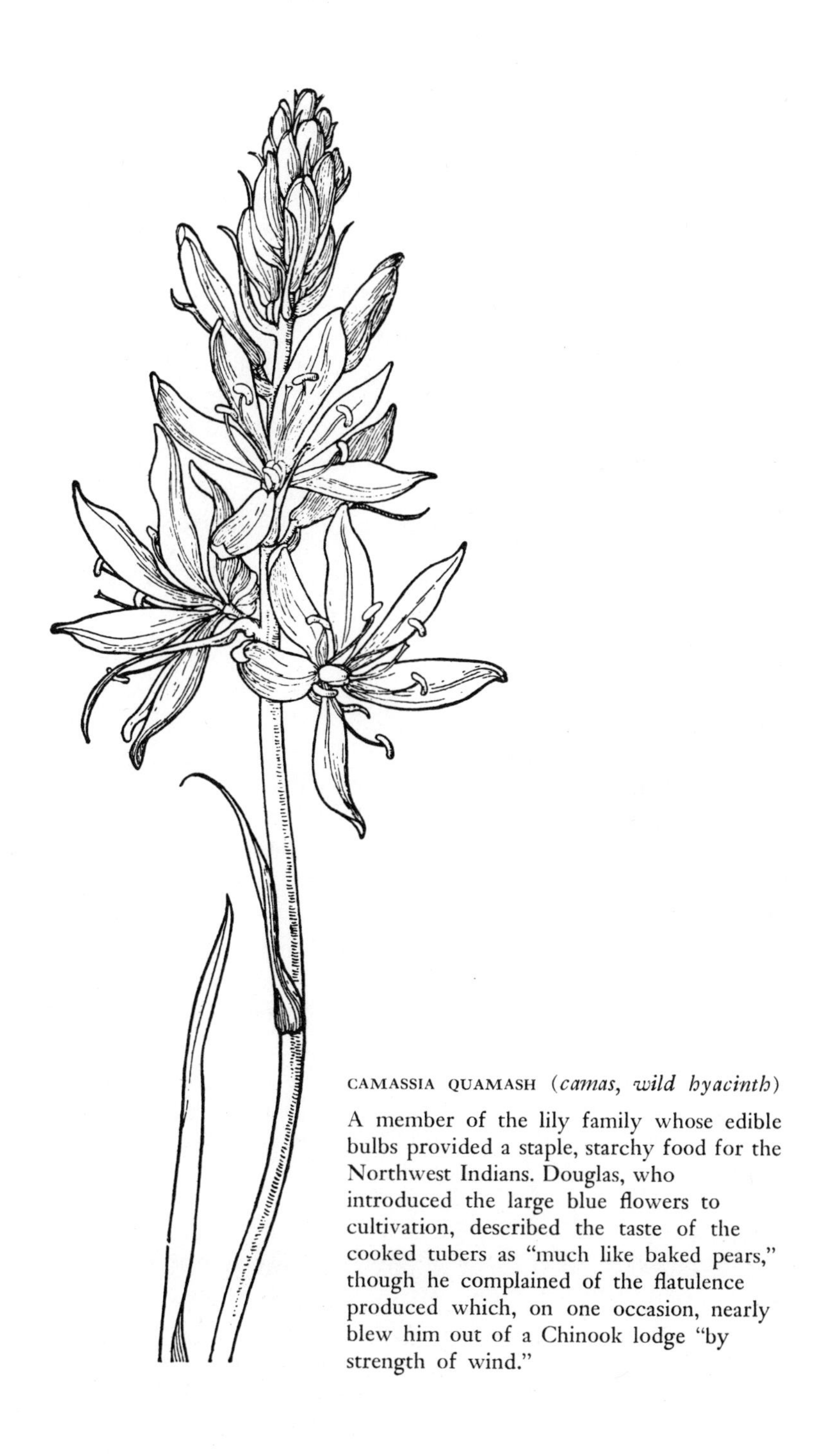

CAMASSIA QUAMASH (*camas, wild hyacinth*)

A member of the lily family whose edible bulbs provided a staple, starchy food for the Northwest Indians. Douglas, who introduced the large blue flowers to cultivation, described the taste of the cooked tubers as "much like baked pears," though he complained of the flatulence produced which, on one occasion, nearly blew him out of a Chinook lodge "by strength of wind."

families would live in such a lodge, each family punching a hole in the roof to permit smoke from its cooking fire to escape. Bunks were apportioned around the walls, unscreened, but nothing went on in them which by tribal convention required screening. William Clark had long ago noted that Indian girls "made sport" in public, an indication that little distinction was made between various bodily functions.

Slaves shared quarters with their masters and ate the same food. As an institution, slavery had no economic basis among the Chinooks but was maintained for prestige reasons. Slaves—originally prisoners of war but, by Douglas's time, more apt to be kidnap victims—were expensive to purchase and a nuisance to maintain, values highly regarded in conspicuous consumption. They were expected to earn their keep by stoking fires and other light chores, but were rarely disciplined for their laziness or misbehavior; in fact, their independence and insolence were causes for frequent complaint. However, though the conditions of bondage were tolerable enough while all went well, there was always the implicit understanding that the slave remained alive only at the will of his master and that he had no redress against burial six feet deep to consecrate a totem pole, or forcible conversion into a spirit in order to accompany a deceased member of the owning family into another world.

Douglas ate many meals in Indian lodges, though possibly with restrained enthusiasm because of the quamash hazard. The main dish was inevitably salmon, inevitably boiled. Stones were heated red hot, then dropped into a pot, chunks of fish being added when the water bubbled. What surprised Douglas at first was that the pots were made of reed fibers so tightly woven as to hold water and tough enough to resist heat. Along with quamash —sometimes roasted separately, but more often tossed into the fish stew— salads were served in spring when greens were tender. In summer and autumn, fruits and berries were eaten fresh and, during the rest of the year, dried.

Hot stones were also used to heat "Indian steaming huts or vapour baths." Douglas was intrigued by this sybaritic luxury: "The bather goes in naked and remains until well steamed; he immediately plunges into some pool or river." Intrigued but resistant, "My curiosity was not so strong as to regale myself with a bath," he tells us.

As Douglas continued to learn about Indians, they became increasingly curious about him. For a long time his occupation baffled them. They could understand what the other King George men were doing on their river—beaver skins had value to them, too. But that a young chief should be sent all this way to collect twigs and flowers, and sometimes rocks and bird skins, didn't make sense. Finally they concluded that King George was mad—as indeed he had been—and that Douglas was collecting novelties to amuse him. They called Douglas the Grass Man and treated him with deference because of his connection with an unhinged mind, always a matter for superstitious awe.

Douglas was aware that a legend was building about him and he employed various means to encourage it. He tells us gleefully about the strategems he pulled. "My canoe-men and guides were much surprised to see me make an effervescent draught and drink it boiling, as they thought. They think there are good and bad spirits, and that I belong to the latter class, in consequence of drinking *boiling* water, lighting my tobacco pipe with my lens and the sun, and they call me *Olla-piska,* which in the Chenook tongue signifies *fire*."

So far so good. We are willing to believe that the hocus-pocus with Epsom salts and a magnifying glass might be good for a gasp and a screech or two. But then he goes too far. "Above all, to place a pair of spectacles on the nose is beyond all their comprehension: they immediately place the hand tight on the mouth, a gesture of dread or astonishment."

No! It's too much to swallow that Indians connected with Fort Vancouver had never before seen eyeglasses. Dr. McLoughlin wore them for reading, and there must have been at least one Company clerk dubbed "les yeux quatres" by the *voyageurs,* the equivalent of Specky Four Eyes. As for the "gesture of dread or astonishment," it's just too theatrical to be real.

We are left with the possibilities that either Douglas concocted the incident to entertain English readers of his journal or—and this seems more likely—his Indian companions egged him into performance for their secret amusement. Instead of Douglas, as he supposed, playing tricks on them, they may well have been having fun at his expense.

What Douglas and his white contemporaries never seemed able to grasp was that they were living among worldly-wise and sophisticated peoples. By any measurement of wealth, social organization, or artistic achievement, the tribes of the lower Columbia constituted the most stable and mature Indian society in North America. Only they, with an assured food base, living free from the threat of hunger, in permanent and substantial villages, could develop painting, sculpture, weaving, carving, and such unique entertainments as tribal dancing and the potlatch to a degree of excellence and prodigality unknown elsewhere.

Theirs was a culture strong enough and self-sufficient enough to resist even the white man's twin corruptions of trading goods and liquor. They were fortunate in that they never seemed to develop the thirst for firewater that consumed tribes east of the Rockies, but they also remained largely apathetic to the bright commodities displayed in Indian Hall. They failed

Dr. McLoughlin as beaver hunters, not through lack of goodwill, but because the rewards offered for hard work were not enough to offset the pleasures of the leisured life. As Douglas had observed, they were "much prejudiced in favour of their own way of living."

In September Douglas returned upriver to the Grand Rapids in order to complete the second movement of a plant collector's cycles. During June he had located the plants that interested him and gathered leaves and flowers to press and dry as specimens. Now he went back to relocate the plants—not always easy in shriveled autumn guise—in order to harvest seeds, bulbs, or roots.

With one of those remarkable bursts of energy that amazed even his vigorous contemporaries, Douglas undertook to climb the heights above the rapids. He did a thorough job of it, scaling the escarpments to both north and south of the river. The first climb, on what is now the Washington State side, took three days, and was called by him "one of the most laborious undertakings I ever experienced." On the south, the Oregon side, the ascent was easier, the "only steep part near the top."

Douglas was certainly the first white man and probably the first man of any description to climb both ridges. He may also be responsible for naming the mountain range in which they occur. He makes no claim to this distinction, but in his journal he refers to the "Cascade Mountains"—the first known instance in which the name of the rapids was given to the range.

In the spring of 1826 Douglas was forced into a decision. His original instructions had been to return to England after a full year of botanizing on the Columbia, but, as 1825 ended, he realized that he had barely begun to explore the enormous territory through which the river flows.

Accordingly he wrote to Secretary Sabine in a humble vein, apologizing to the Society for his decision to stay on, but explaining why it was necessary if justice was to be done to "a country deserving the strictest research." He offered, if expense was considered an objection, to "labour for this year without any remuneration." He also included the hope that "if the motive which induces me to make this arrangement should not be approved of, I beg it may at least be pardoned."

There was, of course, more than a little playacting about all this. Douglas was about as doubtful of official approval as Leonidas had been at Thermopylae when he held back the Persians without orders, or the Dutch boy at the dike when he stuck his finger into the leak without consulting the local burgomaster. But he was entitled to his moment of drama. It was a resolute decision, since he had collected more than enough horticultural, geological, and zoological novelties during his first season on the Columbia to return home without criticism.

Now, in consultation with Dr. McLoughlin, he planned an ambitious schedule for the year 1826. He proposed to leave Fort Vancouver with the first brigade of spring and, from then until the snows of autumn, shuttle between the forts of the Hudson's Bay Company, traveling by its canoes and horses, propelled by its *voyageurs*, and guided by its Indians through the forests and plains of its vast territories, which extended from the Rockies to the ocean and from California to Alaska, a jurisdiction greater than the Holy Roman Empire in its prime, though dedicated to a somewhat different purpose: the pursuit and slaughter of furbearing animals for profit.

5

EMPIRE OF THE BEAVER

Beavers are not unique to the North American continent. They were once widely distributed throughout Europe. The signal for their extermination was the introduction, at about the time of Elizabeth I of England, of a new fashion in men's headgear—the beaver hat.

In its earlier forms, the beaver hat was not the stovepipe structure that might immediately come to mind. Sir Walter Raleigh was possibly wearing a stylish round version when he spread his cloak over that famous mud puddle, and Sir Francis Drake may have been wearing a three-cornered model (still popular in George Washington's day) when the call came to knock off bowls and get ready for the Spanish Armada. By the time Queen Elizabeth passed from the scene in 1603, so had English beavers, hunted to annihilation for the expanding market. They left memorials behind in places called Beverley—literally, "beaver-lea," or glade of beavers.

As the search for beavers widened, the Hudson's Bay Company came into existence, chartered in 1670 to trade in furs throughout the lands draining into the great inland sea of the North American continent. "Furs" included mink, sable, bear, wolf, and fox; but from the beginning, 80 percent of the traffic was in beavers. Not only were they easier to kill, but their pelts kept rising in value as the demand for beaver hats steadily increased.

A new class of mercantile city men emerged in Europe during the eighteenth century—men who found that the beaver hat suggested substance, stability, and sound credit. Later, when the high-crowned, cylindrical topper was introduced, something like mania gripped otherwise rational male populations—interestingly, a mania that cut across social lines. To the aristocrat and upper classes, the beaver top hat symbolized the Age of Elegance—rigid, formal, arrogant. For lesser men, it represented much denied them—gloss, distinction, and a lordly heightening of stature. So by a quirk unique in fashion history, the same hat served for both the duke and his footman, with only styling and quality providing the necessary distinctions.

Hat manufacture was wasteful of beaver skins because only the hair, shaved off with razors, could be used. The notched strands were worked and steamed together into a felt, from which the hat was shaped and molded. No better use was ever found for the shaved pelts than to boil them down for glue. Women's coats, muffs, stoles, and lap robes were also made from beaver skins, but these were comparatively low-profit items.

The newly chartered Hudson's Bay Company set up trading posts in the wilderness, displayed its beads, nails, hatchets, and other trading goods, and invited the Indians to bring furs in for barter. The response was overwhelming. So many beaver skins were turned in that, in 1700, three-quarters of the year's collection was burned to prevent glutting the European market. The sickening smell is said to have lingered for months, while a greasy pall covered everything.

Part of the problem was an upstart French company operating out of Quebec which, by 1715, produced about 100,000 furs annually. The Hudson's Bay Company was now limiting its own harvest to some 150,000 pelts, but the combined total was still too much. Prices continued to fall.

A colonial war bailed out the British company. When General James Wolfe defeated Montcalm at the Battle of Quebec in 1759, French enterprise in Canada ended with French rule, and the Hudson's Bay Company could return to the enjoyment of its former profits, said to be as high as 2,000 percent.

Naturally such tidy returns didn't go unnoticed, and, in 1784, a group of Canadian traders, brushing aside the Hudson's Bay Company's claims to monopoly, entered into competition under the name of the North West Company. The Northwesters, based in Montreal, were a different breed of men from the staid clerks who staffed the trading posts of the Hudson's Bay region. They were adventurous, hard-driving soldiers of fortune who soon extended their fur frontiers north to the Polar sea and west to the Rockies. One of their great men, Alexander Mackenzie, not content with following the river named in his honor to its Arctic mouth, also cut across the mountains and streams of what is now British Columbia to reach the Pacific—the first man ever to cross the continent.

Stimulated by Mackenzie's achievements, other Northwesters continued an epic era of exploration. Simon Fraser followed the river named in *his* honor from the Rockies to its mouth at the present city of Vancouver, British Columbia. David Thompson, three years later in 1811, traced the Columbia for more than twelve hundred miles from its source to its estuary.

Though the Hudson's Bay Company enjoyed a shipping advantage—its principle port on Hudson's Bay was almost at the geographical center of Canada—the Northwesters competed by developing canoe routes through lakes and rivers in order to bring furs to Montreal at comparative cost. Fierce price wars ensued, during which profits, though still large by our standards, were driven down.

Bitter rivalry between the companies extended down through the ranks at all levels. Hudson Bayers found the flamboyant Northwest partners particularly obnoxious. These latter were given to such antics as riding through the streets of Montreal on silver-shod horses, or standing on the steps of Beaver House scattering coins to children. One anecdote concerns a partner who rode into a fashionable restaurant and demanded a bucket of tea for his horse. He grandly left a blank check to pay for breakages to mirrors, tables, and chairs which his steed had kicked in.

Northwesters called Hudson's Bay men "book-keepers." The bookkeepers responded that at least they weren't braggarts, going about thumping themselves on the belly, boasting how much cold they could take. The rivalry progressed from words to warfare in which bullets were fired and men were killed. Peace was finally forced upon the factions by an outraged public and, in 1821, a merger took place. More accurately, the North West Company, weakened by extravagances and a falling market, sold its trading posts, fur stocks, and all other assets to the older organization.

So, in the end, the bookkeepers won out over the "lords of lakes and forests." But there was to be a difference. Many capable North West partners (including Dr. McLoughlin) came over in the merger, revitalizing the management of the Bay Company. Plans went forward to explore new territories, pioneer new routes, and establish new trading posts. Fort Vancouver was built during the first flush of this expansive era, and the spirit of enterprise was still high when David Douglas arrived upon the Columbia in 1825.

On Monday, March 20, 1826, at four o'clock in the afternoon, Douglas left Fort Vancouver for his Grand Tour of the Columbia basin. He left with the Hudson's Bay Company's eastbound express, the annual expedition that rowed and tramped its way across the continent from Pacific to Atlantic. There was also a westbound express, and it was always a great moment when the two met somewhere east of the Rockies. The expresses carried mail, account books, and sometimes supplies for interior trading posts, but never furs. Furs were transmitted back to England by ships which, like the *William and Ann*, called once a year on either side of the continent.

The departure of the express was a ceremonial and rousing occasion. The entire population of the fort assembled on the wharf, women and children dressed in their best, with many Indians ambling in for the excitement. The bagpipers were always out in force, parading in kilts, their uddered instruments wailing eerily.

It was above all else a formal occasion. Dr. McLoughlin wore his best frock coat and a turn-around black silk vest. He carried his gold-headed cane and, of course, a high beaver hat perched on top of his flowing white hair. The *voyageurs*, who looked forward like children to performing in all their finery, were shaved and scrubbed, with their shoulder-length locks braided with ribbons. More ribbons were wound around their foreheads, sometimes clashing with the gaudy colors of shirts and belts. The final fiesta touch was provided by moccasins beaded in every hue of the rainbow.

Douglas made his contribution to the pageantry by wearing a tartan shawl and a Glengarry cap. The shawl may have been woven in the crossed green and purple of the Douglas clan, but not necessarily; we know of at least one other occasion on which he wore the orange, purple, yellow, and black of the Royal Stewarts.*

Brilliant though the costuming might be, it was left for the leader of the express to cut the finest figure. Whether he was a chief trader or a clerk, a ruffled shirt and a satin waistcoat were *de rigueur*. Over this hung a long cloak lined with silk, and, topping all, would tilt the inevitable beaver hat.

There were two boats in the express that afternoon, flat-bottomed, clinker-built, tapered at each end. The *voyageurs* sat with paddles poised, waiting for the signal to take off. The leader, Chief Trader John McLeod, doffed his hat to Dr. McLoughlin, requesting permission to depart. The doctor responded by calling "Godspeed!" and the cannons of the fort fired.

The *voyageurs* took it from there, bursting into song as their paddles struck the water. Fresh and vigorous, showing off with every stroke, they

* Tartan patterns, the invention of Scottish weavers, were not considered the exclusive property of individual clans until their adoption as regimental insignia by the British military about the middle of the nineteenth century. A painting as late as 1745 shows a Scottish warrior authentically clad in three different tartans, one for his cloak, another for his kilts, and the third for his socks.

literally rammed the boats upriver. Their songs trailed back, growing fainter as the flotilla passed around a bend and was lost to sight.

"We camped . . . a few miles from the establishment, at dusk," Douglas records in his journal. Departures were timed for an early stop on the first night. Out of sight of the fort, work clothes appeared and finery was tucked away until arrival at the next Company post, when a halt would be called to dress up again.

There was trouble at the Dalles. The area was inhabited by pirating Indians who preyed on river traffic. Alone among the tribes of the lower Columbia, the Wishrams lived by violence, or the threat of violence, exacting tribute from boat crews forced to portage around the rapids and from fishermen netting salmon from "their" rocks.

To avoid friction, Dr. McLoughlin had accepted the policy of paying tribute, so much tobacco per boat; but on this occasion, after permitting the portage to proceed all the way to smooth water, the Wishrams arbitrarily increased the toll price and refused to allow departure until the higher levy was paid. They may have been interested in a cargo of calves and pigs the express was carrying, or they could have been emboldened by visiting kinsmen—Douglas counted "at least 450 savages." In any case, when their demands for more tobacco were refused, they prepared to attack.

Chief Trader McLeod, standing in front of the boats, pushed back the first onrushing Indian. His action triggered a sequence of events that might have come straight from the Leatherstocking tales or from annals of western adventure. As Douglas tells it: "A fellow immediately pulled from his quiver a bow and a handful of arrows, and presented it at Mr. McLeod. As I was standing on the outside, I perceived it, and, as no time was to be lost, I instantly slipped the cover off my gun, which at the time was charged with buckshot, and presented it at him, and invited him to fire his arrow, and then I should certainly shoot him."

Into this tableau of frozen suspense stepped a heroic figure, a six-foot-six chief of a tribe of Cayuse Indians "who are the terror of all other tribes west of the mountains." This particular chief seemed also endowed with statesmanlike qualities because, in the words of Douglas, he "settled the matter in a few words without any further trouble."

The express boats pushed off, the Cayuse chief accompanying them until they had reached a safe distance. He was "remunerated by Mr. McLeod" with gifts of tobacco, and by Douglas with a silver shilling to hang from his nose.

Fort Walla Walla, situated at the first great bend of the Columbia, where it button-hooks north to become, in terms of today's geography, the exclusive property of the state of Washington, was surrounded by the same brown and arid desert that Douglas had seen begin over a hundred miles to the west. It was a poor country for botanizing and even worse for beavers, without a tree in sight to supply them with food or building materials. The trading post had been established in the hope that Indians would bring skins down from the neighboring beaver-filled mountains, but it proved a futile hope. Here, as elsewhere, the Indians proved unambitious, or "unwilling to exert themselves," as the phrase most often went.

In command of the adobe stronghold at Fort Walla Walla was Chief Trader Samuel Black, then in his early forties. He was a man of "enormous stature" and has also been described as a "Don Quixote in appearance, ghastly, raw-boned, lantern jawed, yet strong, vigorous and active." Born illegitimately in Aberdeen, Scotland, and completely without formal education, he had taught himself to find his way around the classics and also possessed an amateur's enthusiasm for geology. Like Dr. McLoughlin, he had originally been a Northwester and had transferred after the merger; but unlike the doctor, he hadn't converted well. Spiritually, he was still a *coureur des bois,* a runner in the woods of his youth.

Perhaps it was a mutual interest in geology that drew the dour chief trader and the young collector together. They climbed some of the rocky mounds in the vicinity of the trading post, from which Douglas saw "to the south-east, at a distance of ninety miles . . . a ridge of high snowy mountains." These are the Blue Mountains, and Douglas was cordially invited to return later in the summer to explore them, making Fort Walla Walla his headquarters.

The express moved on again upriver. Douglas scrambled along the shore, botanizing, until he exhausted himself keeping up with the boats; then he'd be hauled aboard to doze under a blanket, lulled by the *voyageurs*' rhythmic songs.

The days passed, days that started before dawn so that the boats might be well under way by daylight, with only two formal stops thereafter, for breakfast and the midday meal. The day's run ended at dusk, allowing just enough time to pitch tents, start fires, and feed the stock before darkness fell. The *voyageurs* might continue singing long after supper, but Douglas, after checking his specimens and recollecting his day for his journal, was soon asleep.

There were portages—at Priest's Rapids, Wanapum, and Rocky Reach—stretches of picturesque, fast-moving water through deeply carved canyons. Needless to say, all these rapids and others along the same stretch of river have long since been drowned under the backed-up water from dams.

The express put ashore for a few days' rest at the trading post of Fort Okanagan * where the route changed direction, moving upriver in a sharp bend to the east. Then, fortified by fresh rations, it set off again on a twisting, turning course through a still bleak countryside.

On Tuesday, April 10, the boats turned north once more and Douglas was gladdened by a sight he was beginning to think had vanished forever—a green landscape. With dramatic suddenness, as abruptly as it had stopped four hundred miles downriver, the forest began again. They were back in a climate of soft air and captured rain clouds.

"This part of the Columbia is by far the most beautiful that I have seen," Douglas writes, with perhaps exaggerated appreciation after the long drought. "Very varied, extensive plains, with groups of pine-trees, like an English lawn, with rising bluffs or little eminences covered with small brushwood."

It is a measure of the achievement of eighteenth-century English landscape architects that travelers, for a hundred years afterward, kept seeing compositions in the wilderness that reminded them of the artificial lakes, hills and rolling meadows of Blenheim, Stowe, Rousham, and other famous estates laid out in the "natural" style. It was not only David Douglas on the upper Columbia; thirty-four years earlier, Dr. Archibald Menzies had seen on the shores of Puget Sound "a rich lawn . . . here and there adorned by aged pines with wide-spreading horizontal boughs . . . the whole seeming as if it had been laid out from the premeditated plan of a judicious designer."

Douglas and Menzies knew a good deal about gardens, especially Douglas who had trained in them. But George Simpson, field governor of the Hudson's Bay Company, certainly had no such expertise. He was a practical man who prided himself on having no interests outside of business; yet he described the site he selected for Fort Vancouver in the following glowing words: "The finest plain . . . watered by two very pretty small Lakes and stu[dd]ed as if artificially by clumps of Fine timber . . . I have rarely seen a [finer] Gentleman's seat in England."

It was as if, in that disciplined age, undisciplined nature was considered too chaotic to be admired for itself. It could only win full approval when, by accident, it assembled elements into a design that might have been created by human agency. Or, to reverse Shakespeare's aphorism, nature was at its best when it held the mirror up to man.

* Spelled Okanogan in the United States; but Canada still retains Douglas's spelling, pronounced Oaken-ahken.

Near the junction of the Columbia and Spokane rivers, the express found another Hudson's Bay party encamped and waiting for it. This was a contingent under the command of Chief Trader John Warren Dease, on its way to found a new trading post at Kettle Falls. It was an advance guard to take delivery of the calves, pigs, and other supplies the express brought from Fort Vancouver; wives, children, and operating personnel were still at Spokane House, the old post which was being abandoned for the usual reason—a lack of beavers.

The transfer of livestock completed, there was a parting of the ways: Chief Trader John McLeod moved on with the express for his long journey across the continent; Douglas, joining the Dease flotilla, proceeded upstream at a more leisurely pace.

Douglas had looked forward to meeting his new host, if for no other reason than the magic of his name. For John Warren Dease was a brother of Peter Warren Dease, assigned by the Hudson's Bay Company to help the Franklin expedition reconnoiter routes to the Arctic. Douglas, always prone to hero worship of explorers, reserved special niches in his pantheon for Captain John Franklin and everyone associated with him in his hazardous enterprise. He had been flattered to meet in London Dr. John Richardson, physician and naturalist of the Franklin party, and he now was prepared to lavish respect on John Warren Dease as the brother of another hero, except that circumstances made it difficult. For his new companion was an alcoholic. Douglas couldn't have been with him long before he realized that the cups of tea the chief trader sipped continuously through the day were heavily freighted with brandy. Dease's voice, strong in the morning, sank to a whisper in the afternoon and he usually had to be helped from the boat when camp was made.

The infidelities of his Indian wife were popularly supposed to have driven Dease to the bottle, but almost certainly it went deeper than that. He seems to have been of a sensitive, even poetic nature. Of all Douglas's hosts on the river, Dease took most interest in his work, often accompanying him on botanical walks and sharing his enthusiasm in his discoveries. In contrast, he seemed increasingly distressed by his own profession. The killing of beavers preyed upon his mind. He lost his appetite for their flesh and became unable even to watch their skins being brought in for barter. He was aware, as his weaknesses became known, that he had no future in the Company, yet seemed incapable of summoning up the resolution to leave it. His conflicts, unresolvable, were inevitably leading to a self-destruction that surprised no one when it came.

The boats reached Kettle Falls after three days of steady pulling against the current. While the *voyageurs* portaged supplies and equipment to the "small circular plain" above high-water mark where the new Fort Colville was to be built, Douglas moved up the river bank to examine the falls. He described the turbulent waters as a double cataract, the Kettle River first pitching down some twenty-five feet to join the Columbia, then the combined waters charging together in a second spectacular plunge.

Contrary opinions exist as to how the "Kettle" name originated. One version has it deriving from the potholes gouged out of the rock by the swirling water, the other, from the fiber-woven cooking pots sold at the site. David Thompson, descending the river for the first time, remarked on the hundreds of Indians from many tribes who gathered at the falls to barter baskets for the thousands of salmon caught in the cataracts. He leaned toward the cooking-pot theory, but we will never know for sure. The area is now cut off to archeologists and anthropologists, as to everyone else, by the backed up waters of a dam—the Grand Coulee—a hundred and fifty miles downstream. The falls are submerged, all thirty-three feet of them; so is Fort Colville, built three-quarters of a mile from the original river bank on high ground. The eellike Franklin D. Roosevelt Lake obliterates both scenery and history beneath its placid waters.

Through April and May, Douglas botanized on foot and horseback through what is now northern and eastern Washington. Among many interesting plants was one notable discovery: a pine with bark cracked like an alligator's hide. Oddly, Douglas seemed unaware of an important find. There were none of the hints such as "worthy of strict attention" with which he often signaled a putative winner. He simply sent along specimens and seeds of a particularly resinous conifer that turned out to be the western yellow pine *(Pinus ponderosa)*, the most important tree in lumbering after the Douglas fir.

By June he was back at Fort Walla Walla to take up Chief Trader Samuel Black's offer to outfit him for an expedition into the Blue Mountains. One night in the adobe compound he unexpectedly collected a specimen for the Zoological Society. He was asleep and unaware when a family of pack rats entered his room and carried off his razor and shaving brush; but he was awakened when a blundering rodent returned for further thievery. The journal describes what happened next: "As he was in the act of depriving me of my inkstand . . . I lifted my gun (which is my night companion as well as day, and lies generally alongside of me, the muzzle to my feet) and gave him the contents."

The shot, echoing through the building, brought Samuel Black running. When he found out what had happened, he shook with laughter. Then he launched into reminiscences about pack rats he had known along the Peace and Mackenzie rivers. In winter, anything bright and shining that wasn't nailed down disappeared.

Talking about the old days revived memories for Black—in particular, memories of the conflict between the fur companies that had absorbed his mind and energies for fifteen years. As a young Northwester, he had seen the Hudson's Bay Company's early efforts to invade the Athabaska territory and had helped plan the strategy that drove them out. While his superiors bargained with the Indians for the exclusive purchase of all furs, Black undertook to harass the enemy. By staging fake Indian attacks on their trading post, stampeding horses, exploding mines, and firing the forest, he made business by day and sleep by night virtually impossible. Finally the Hudson's Bay Company, its personnel exhausted and unable to obtain pelts, gave up and withdrew from the Athabaska region.

Black, by now a legend among Northwesters, was made a full partner of the Company and put in charge of resisting all further Hudson's Bay encroachments. His crowning escapade was "taking prisoner" a Hudson's Bay chief factor and shipping him under guard "for trial" at Beaver House in Montreal. Of course the captive was soon released, but the Hudson's Bay Company brought kidnap charges against Black. The legal proceedings accomplished nothing but to give all Canada a hearty laugh. It was probably this exploit that accounted for the ring Black was given by his colleagues, engraved "To the most worthy of the worthy Northwesters."

When the merger came, Black was as reluctant to join the Hudson's Bay Company as the conservatives of that organization were to have him. But George Simpson, the new field governor, insisted. He knew that, without Black, old Northwester wounds would be longer in healing, and he was anxious to get back to profitable fur trading. So he persuaded Black to accept appointment as a chief trader and sent him to the sun-baked waste of Fort Walla Walla. Whether this was done to chasten him for his former misdeeds or in the hope that he could work miracles in the beaverless land was never made explicit.

Later, Black was transferred to a more promising post and promoted to chief factor. But it is doubtful if even this reconciled him to the Bay Company. He wasn't a man who could forgive and forget easily, or make friends with old enemies for the sake of expedience. It always seemed as if he had left the best part of himself in the Athabaska territory, where Black Lake still honors his name, and where he won so many victories that somehow terminated in defeat.

In mid-June, Douglas set out for the Blue Mountains in command of a small expedition consisting of a guide, an interpreter, horses, camping equipment, and provisions for ten days. He had, of course, been nominally in charge of the guides and Indians provided by Dr. McLoughlin at Fort Vancouver and by Chief Trader Dease at Kettle Falls; but, always before, his trips had been to known destinations. The journey into the Blue Mountains was a different matter. It was unexplored territory, generally shunned by white men and Indians alike because of the warlike reputation of the Cayuse tribes. Douglas must have felt some of the authority and purpose he attributed to heroes like Captain John Franklin as he led his contingent across the plains from Fort Walla Walla.

By the third day they were well into the pine-forested mountains and had pitched camp in an alpine meadow. Douglas, driven by his old urge to climb, ascended a peak which he described as being wreathed in "eternal snow" and calculated to be some nine thousand feet in height. He always had trouble with his heights; later he was to create geographical confusion for a hundred years by miscalculating the height of two peaks in the Rockies, and that day in the Blue Mountains he was also wide of the mark. There isn't a summit in the entire range above sixty-two hundred feet.

He did better botanizing. One of the things that made Douglas a great collector was his persistence in tracking down all possible variations of a plant. For example, pentstemons: most collectors would consider their duty done after gathering half a dozen kinds, but Douglas kept looking and wound up with a total of eighteen. He was even more prolific with lupines, finding twenty-one distinct forms in all.

Though the principles of genetics were not known in Douglas's time, it was common knowledge that cross-fertilization could produce new varieties often more spectacular in leaf or blossom than the parent plants. As a practical gardener, Douglas also knew that the more related types available for hybridization, the more startling the results that might be achieved.

Parenthetically, it should be observed that this manipulation of nature made many gardening gentlemen philosophically uneasy, especially if they happened also to be clergymen. It seemed to go against the established dogma of immutability of species given, as we have seen, biblical authority in Genesis. But though the implications troubled some religiously inclined people, others, no less devout, found comfort in the thought that man-made varieties were only artificial and temporary; once out of fashion or "escaped" from cultivation, they rapidly retrogressed into former aboriginal or "natural" states. Strictly speaking, this was not true—no form of life can ever revert exactly to a preceding form—but belief in an ultimate return to a kind of

PENTSTEMON HETEROPHYLLUS (*azure beard-tongue*)

One of the eighteen species of pentstemons that Douglas either botanically discovered or introduced to horticulture. Most are from the Columbia River region, but this one is from California, where it is a denizen of dry hillsides in the Coast Range. The common name derives from the beardlike lower lip of the flower.

PAEONIA BROWNII (*wild peony*)

Only peony ever found on the North American continent; discovered by Douglas in 1826 in the Blue Mountains of Oregon and named by him in honor of Robert Brown, eminent British botanist.

eternal Eden allowed many a pre-Darwinian conscience to sleep in peace.

Somewhere in the lush valleys of the Blue Mountains Douglas found a purple and yellow peony, an epic strike since it is the only peony type ever to be found on the North American continent. He realized he had a big one and named it big, *Paeonia Brownii*, in honor of Robert Brown, curator of Sir Joseph Banks's herbarium and widely celebrated as the greatest botanist of his time.

He made more discoveries in subsequent days of botanizing. He named a flowering currant *Ribes Munroi* after Donald Munro, head gardener at Chiswick, and a yellow lupine, which reminded him of Scotch broom, after John Turner, an undersecretary in the Horticultural Society's front office. This latter nomination caused a flurry when it reached London because that year Turner had been exposed as an embezzler and was still being hunted on the continent. When Douglas returned to England, the predicament was explained to him and he obligingly renamed the plant *Lupinus Sabinii.*

Exhilarated by his discoveries, Douglas prepared to move farther into the mountains. He had heard from the *voyageurs* about a beautiful round valley which they called *La Grande Ronde*, the Great Circle or, as they seemed to imply when they spoke of it, the Magic Circle. A dozen years before, a hunting party, lost and starving, had stumbled into it and remembered it ever afterward as full of sunlight, birds, and flowers. Friendly Indians had guided them back to the river, but none could forget the beautiful place. The *voyageurs* were a bit vague as to the location of *La Grande Ronde*, but Douglas meant to press on until he found it.

Then, unexpectedly, his guide and interpreter, a twelve-year-old French-Canadian, decided that it was time to return home. The boy informed Douglas that there were hostile Indians ahead and he was afraid to go farther. Douglas insisted that they must proceed in spite of risk, and instructed that the guide be so informed. The interpreter launched a spirited harangue to convince the guide he must do his duty, but failed. The Indian was adamant.

Douglas feared that the "rascally boy told the Indian the reverse of what I wished him to," but saw no way out of his dilemma "as it would be very improper to force [the guide] to go with me, and impossible for me to go alone." He was "reluctantly obliged . . . to give up."

They turned back, Douglas aware that he had failed in a test of leadership since leaders somehow persuade or bully men to follow them. The matter might be considered unimportant—leadership is not a primary requisite for plant collecting—except for the opinion Samuel Black may have formed when the party straggled back prematurely to Fort Walla Walla. His immediate reaction was sympathetic. He reoutfitted Douglas and sent him off again to the Blue Mountains with more reliable companions. But the memory of that failed expedition may still have lingered when, years later, Douglas was pleading for a final chance to prove himself.

Douglas skirted the mountains to the north on his second trip, hoping to find a stream that might lead him through to *La Grande Ronde*, but he was disappointed. The botanizing was poor too, so, after a couple of rather aimless days, he headed back.

Strong westerly winds developed, blowing sand in the men's faces as they rode toward the fort. Douglas was unable to protect his eyes against the constant barrage of grit and reached the trading post "much worn down and suffering great pain from violent inflamation of the eyes." This entry was made on June 29, and he adds: "To read or write I cannot . . . without pain." By July 4, however, he was "drying the plants collected, gleaning and putting up seeds," and writing voluminously in his journal.

This satisfactory progress suggests adequate medication and nursing. The medication was probably a boric acid solution. Sore eyes were common in that sandstorm area and the traditional salve would be available. But an inflammation as acute as Douglas describes must have received more than superficial care to reduce it so rapidly. Hours of lying in a darkened room are indicated, with constant changing of solution-soaked bandages.

Who was the devoted nurse? Samuel Black? Hardly; and that leaves only one person with the skill and patience required. Douglas doesn't tell us about her, but then he is always cautious about mentioning women in his journal. Occasionally, during his visit to the United States and, later, in California, we catch glimpses of wives and daughters, but never on the Columbia. That was an all-male society, at least in myth, and Douglas worked overtime to support the fiction. He remained deaf to women's laughter behind closed doors in trading posts up and down the river, and it would have been the same at Fort Walla Walla but for the accident of sand in his eyes.

He probably called her Mrs. Black, though her name was Angelique Cameron. She was either the widow or daughter of a Scottish trader and of half-Indian descent. No one knows where Black met her, but they were together when he assumed command at Fort Walla Walla; and no one knows why, after so many years and four children, they never married. All that's certain is that Black cared for her deeply and when, fifteen years later, he staggered into her room dying from the Indian's bullet, his only thought was to make a will to provide for her and the children.

To Douglas, lying in the dark, Angelique Cameron's hands against his face must have reminded him of other hands, her presence in the room of another presence. During the long hours in that July heat, he could hardly help but think back to the previous July, to the village by the sandy-bottomed river, to Cockqua's lodge.

He thought back, but he must also have tried to push the thoughts aside, because it was all over now. She had understood why, or he hoped she had. In any case, there was no going back.

But perhaps he found that resolution dissolving even as he restated it. She had unleashed too many phantoms in his mind. He knew, he must have known, that he would see her again before he left for England.

6

PATTERN OF A HERO

DOUGLAS RETURNED TO FORT VANcouver at the beginning of September. It was earlier than he had intended—too early for many autumn-maturing seeds—but he came plummeting down the river, covering some six hundred and fifty miles in twelve days, because he had received word that a brigade was about to depart for the mountains of the south under the command of Chief Trader Alexander Roderick McLeod.

We've observed that Douglas was predisposed to admire men of action, particularly explorers. Within a year he was to meet his archetypal hero, Captain John Franklin, prober of the Arctic, a gallant if somewhat lame-brained naval officer of whom it was to be said—though never by Douglas—that he "won undying fame through failure." Within two years, when Douglas would be something of a celebrity himself, he was to meet notables connected with other publicized expeditions. But it is doubtful that any of them made more of an impact on him than had the Hudson's Bay Company's Alexander Roderick McLeod. There was a style to the man. His open manner and easy laugh evoked confidence, while his occasional frowns suggested depths of character. Above all, he seemed always alert and poised for the adventure suggested by his dashing name.

Douglas first saw the chief trader under the stirring circumstances of a brigade arrival. "At midday on the 18th (November, 1825) the [e]xpress, consisting of two boats and forty men, arrived from Hudson's Bay which they left on the 21st of July," he notes in his journal. "They were observed at a distance of some miles, rapidly descending the river." He "hastened to the landing place" where, to the skirl of bagpipes and cheers of Fort Vancouver's assembled population, he saw McLeod standing tall in the lead boat, every inch the commander of the expedition. As the Fort's cannons fired, off swept McLeod's high beaver with a flourish. The chief trader, then forty-five years of age, handsome, tall, vigorous, unquestionably made a lasting impression on Douglas even before their formal introduction.

It is probable that the senior officers at Fort Vancouver, especially those who had worked with him before, viewed McLeod's arrival with less than enthusiasm. For despite his engaging personality, the chief trader had a reputation for blundering, and now it was rumored that he'd upset the pickles again, that his transfer to the Columbia department was not the promotion he supposed but a ruse to get him out of the Mackenzie River country before he floundered into disaster.

McLeod seemed completely unaware that he was in any kind of official disfavor. He was still full of the project that had caused his ouster from his former post, and eager to talk about it, but, as he found his colleagues disinclined to listen, his audience finally narrowed down to David Douglas who, no more attuned to political undercurrents than McLeod, was delighted to be singled out for attention.

The chief trader's story was certainly compelling. He told Douglas that in the region far to the north, from which he had just come, Indians had informed him of "a river nearly equal to the McKenzie [Mackenzie], to the westward of it, running parallel with it, and falls into the sea near Icy Cape.* At the mouth of the said river there is a trading establishment on a woody island, where ships come in the summer. The people have large beards and are very wicked; they have hanged several of the natives to the rigging and have ever since been in much disrepute."

McLeod would have been more interested in the prospect of a great undiscovered river than in the grievances of Indians. Like Samuel Black he was an old Northwester and his mind was filled with glories of other days. He had been twenty-eight years old when Simon Fraser rode his turbulent river down to the ocean, thirty-one when David Thompson navigated the Columbia from its source to the sea. He'd waited his turn for a similar epic expedition, but time had passed and the country seemed to have run out of new great rivers.

Then, belatedly, came this opportunity. He had to take it, but he

* Factually, no such river exists. The Yukon, which alone would qualify in size, rises west of the Mackenzie but flows away from it in a southwesterly direction to the Bering Sea. Streams that emerge near Icy Cape are insignificant.

moved with caution. The time had passed when a field officer could take off in any direction at will. Hudson's Bay men moved only on orders, and orders were not likely to be issued for an enterprise that could bring about a head-on clash with the Russians, unquestionably the wicked men with large beards. So McLeod resorted to guile. He began organizing a beaver-trapping expedition along the unnamed river, as though completely unaware that it might flow through territories claimed by the Czar.

But guile wasn't McLeod's strong suit. Soon enough his real intentions were signaled to Hudson's Bay headquarters, causing Governor Simpson to enter in his secret diary a black mark against him for "preposterous and galling use of authority." McLeod was peremptorily relieved of duty on the MacKenzie and shuttled two thousand miles across country in charge of the express bound for Fort Vancouver. Incredibly, he was so little chagrined at having been kicked across a continent, or so little aware of it, that he could boast to Douglas about "visiting the Polar Sea, [and] the Atlantic and Pacific Oceans, in the short space of eleven months."

As it turned out, McLeod arrived at Fort Vancouver at an opportune time. Dr. McLoughlin had an assignment waiting for a man of the chief trader's temperament—as a brigade commander in the Company's drive to destroy every beaver south of the Columbia River.

This drastic policy had been decided upon to thwart American trappers from across the Rockies. For years the Continental Divide had served as the informal boundary between the Hudson's Bay Company and American fur interests, but increased hunting pressure along the Missouri and Arkansas river systems was driving Yankee trappers across the ridge and down the western slopes of the mountains into what is now the state of Oregon. To turn back these invaders, the Hudson's Bay Company had undertaken to reduce the entire territory, from snow line to ocean, to a fur desert.

All through the summer of 1826, while Douglas botanized on the upper Columbia, brigades from Fort Vancouver, A. R. McLeod's men among them, ravaged up and down streams and rivers, trapping, killing, destroying, leaving behind a beaver wasteland. And through crisp autumn days, when the valleys were decked with the copper, gold, and vermilion of turning leaves and the greenery of conifers sparkled bright on mornings of hoarfrost, the slaughter continued.

Douglas arrived back in Fort Vancouver in time to join McLeod's winter expedition. He was delighted not only to resume his association with a man he admired but also because, in the Umpqua Mountains toward which the brigade was headed, he hoped to find a pine which promised to be not only the colossus but the oddity of its species.

For over a year he'd been tantalized by piecemeal evidence of the existence of this enormous tree. He had seen and eaten the outsized seeds, roasted as a dessert by the tribes of the Multnomah (now called Willamette *) River valley. By offering rewards, he had also obtained twigs, cones (up to a startling 16½ inches in length), and sections of wood exuding a peculiar resin, so peculiar, in fact, that when he wrote to Professor Hooker about it, he was apologetic: "Trees of this *Pinus* . . . produce a substance which, I am almost afraid to say, *is sugar*."

It was a rigid law of botany (and still is) that only broadleaf trees—maples, magnolias, willows, and the rest—run sap that can translate into sugar. Conifers produce gummy, sticky resins suitable only for turpentines. Therefore, what kind of travesty on nature was a pine which, in Douglas's words, "yields a large quantity of resin . . . use[d] in seasoning, in the same manner as sugar in civilized society"? Conclusions would have to wait until he beheld the tree with his own eyes and collected on-the-spot scientific evidence of its peculiarities.

Since the brigade to the Umpqua country was to travel overland, Dr. McLoughlin lent Douglas one of his own horses. Douglas was flattered, but also awed by the responsibility. From the care he lavished on the animal throughout the trip, it is clear that he considered breaking his own leg would be less of a calamity than breaking one of the horse's legs. Douglas always had a poor boy's exaggerated respect for property. Years later, thrashing about in the rapids of the Fraser River after a canoe accident, he almost lost his life trying to save some bulky scientific instruments which weren't fully paid for.

He was ferried across the river from Fort Vancouver and, a couple of days later, caught up with McLeod's brigade as it was assembling personnel and supplies at Multnomah Falls, the present site of Oregon City. It was a territory that, because of its propinquity to Fort Vancouver, had been heavily hit by the search-and-destroy policy. A year previously, Douglas, botanizing through the area, had observed; "This at one time was looked on as the finest place for hunting west of the Rocky Mountains. The beaver now is scarce; none alive came under my notice. I was much gratified in viewing the deserted lodges and dams of that wise economist."

"Gratified" seems a strangely inappropriate word considering that the lodges were only "deserted" and exposed to view because of the dynamite that had blasted the dams and drained the ponds. Some Company officers, inured as they were to killing beaver, were squeamish about duty south of the Columbia. In other regions, farsighted and humane conserva-

* The word "Willamette" (rhymes with damn-it), despite its Frenchified look, is authentically Indian, meaning a "Big river, almost, not quite" in the Kalapooia language.

ERYTHRONIUM GRANDIFLORUM (*dogtooth violet*)

An example of a misnamed plant, at least in its common name. It is not a violet, it is a lily, and while the "dogtooth" description of the corm might apply to the diminutive eastern species, it hardly does justice to Pacific coast bulbs that produce lily stalks two feet high with as many as sixteen flowers on each. The preferred name for the giants first found by Douglas along the Willamette River is fawn lily.

tionist policies were still in effect. Beaver populations were maintained by a refusal to barter for skins during breeding season and by discouraging the trapping of immature animals at all times. Only in the Oregon country were the rules different.

Traders, generally, respected beavers, thinking of them as clean, hard-working, and intelligent animals. Perhaps no white man ever felt the affinity of some Indian tribes who called the beaver "brother" and apologized to its spirit when forced by hunger to kill it,* but most trappers retained engaging memories of the animals, such as the lullaby of a mother beaver singing to her babies on a warm spring night while Indian villagers sat quietly at the pondside listening. The memories persisted during those extermination campaigns even as adult beavers, trying to repair dynamited dams, were bludgeoned to death, while their pups, sliding around in the mud, whimpered like children.

In the end it really didn't matter. When the Oregon Trail was opened up and settlers by the thousands poured into the Willamette Valley, the animals were doomed anyway. Beavers and farmers cannot coexist; their ecological needs are too contradictory.

The Marquis of Bute proved this indirectly when he imported two pairs of beaver and turned them loose on his Scottish estate. At first all went well; the immigrants found a stream to their liking, obstructed it with a dam seventy-eight yards long and went to work on a palatial lodge in the middle of the resulting lake. The only trouble was that they were felling so many trees that the Marquis was forced to limit their activities by a wall in order to save his ancestral park.

From then on it was all downhill. The animals felled the remaining trees and shrubs within their enclosure, then went into a decline. Fresh willow and poplar branches, their favorite foods, were cut and hauled to them daily but they took little interest in what they seemed to consider an artificial diet. They became disoriented and began swimming in broad daylight, unusual for beavers. Shortly thereafter they sickened and died.

In time, the Marquis's beavers, and beavers everywhere, were to obtain their revenge. During David Douglas's first summer on the Columbia, in 1825, a new style in male headgear was introduced in Paris. It was called the Florentine silk hat. It didn't catch on overnight; men's fashions—then, anyway—changed slowly. But within a decade, the high beaver, the very rock on which fur empires were founded, was on its way out.

Today the hat is extinct, but beavers, in the depths of mountain glens and hidden valleys, have made a surprising comeback.

* Among such tribes the early missionaries got short shrift when they refused to ascribe "souls" to beavers.

McLeod's brigade moved slowly up the Multnomah River. It totaled about thirty persons including, in addition to Hudson's Bay Company personnel, Indians, freemen,* independent trappers,* and their families. The Company encouraged, even pressured, men to take their wives and children along on these extended expeditions. It cut down on woman trouble at villages along the way and also prevented men from deserting because of homesickness or when faced by hardships or danger. But though effective in holding brigades together, family encumbrances drastically slowed travel, limiting progress to a dozen or fewer miles per day.

Douglas rode at the post of honor, beside McLeod at the head of the column. The rich vegetation, studded with stands of giant Douglas firs, composed itself into an endless series of green curtains before them, necessitating the cutting and trampling out of a path that could be followed by the plodding pedestrians. They were also on the alert for game, as were other mounted hunters ranging on either side, since the contents of the cooking pots that night would depend on what the rifles brought down during the day. Deer were plentiful but the rank cover made it difficult to get a clean shot at them. A couple of times during the first few days McLeod commended Douglas for his marksmanship, compliments duly noted in the journal.

Gradually the character of the woods changed. They entered a more open country; "fine rich soil," Douglas recorded, "oaks more abundant, and pines scarcer and more diminutive in growth." Because of better visibility, deer were easier to kill and the brigade feasted.

Abruptly they passed into a new scene—miles of blackened tree trunks and rubbish caused by a recent fire. Forest clearance *via* the torch was an old Indian custom, carried on for thousands of years. The fires burned out undergrowth and tree canopies, admitting sunlight and encouraging meadows of grass; the grass lured the deer into the open where they could be killed by bows and arrows. Clover, flourishing among the grasses, attracted honeybees and bears. The Indians didn't care which; they were partial to both. Sometimes, of course, fires raged unexpectedly long with too-extensive devastation. There is evidence from tree rings of one blaze around the year 1690 that consumed half a million acres.

The blackened waste temporarily put an end to Douglas's botanizing. It also raised problems about feeding the horses, since no green yet showed through the charred stubble. To save undue wear and tear on Dr. McLoughlin's steed, Douglas walked all day across the ashes, to learn that they concealed rocks that stubbed and bruised his feet.

* Freemen were *voyageurs* and laborers who had worked out their contracts as employees but stayed on with the Company as trappers. Independent trappers were often Americans, down on their luck, willing to go shares on furs in return for subsistence.

Greenery finally returned and, with it, adventure. One of the hunters shot at a grizzly, failed to bring it down, and was treed by the wounded and enraged beast. It was a close call. "The bear caught with one paw under the right arm," Douglas tells us, "and the [other paw] on his back. Very fortunately his clothing was not strong, or he must have perished." As it was, he remained half-naked up a tree until his shouts brought help.

Most of the cooking in camp consisted of unimaginative venison stews, but there could be pleasant surprises. One of the French-Canadian hunters turned out unexpectedly to be quite a chef. He "roasted a shoulder of . . . doe for breakfast," Douglas records, "with an infusion of *Mentha borealis* [a species of mint] sweetened with a small portion of sugar . . . the meal laid on the clean mossy foliage of *Gaultheria shallon* in lieu of a plate."

They were now approaching the southern end of the Multnomah Valley, in the vicinity of the present city of Eugene. On a Sunday—"known only by the people changing their linen"—he climbed a hill and found the first chestnut he had seen in the west, the spectacular giant chinquapin. "A princely tree," Douglas rightly observes, "60 to 100 feet high, 3 to 5 feet in diameter, evergreen, the leaves having a dark rich glossiness on the upper surface and rich golden-yellow below."

Shortly afterward, following a spicy scent, he encountered the California bay *(Umbellularia californica)*, a handsome roundheaded evergreen, the "leaves very aromatic . . . and when rubbed in the hands produces sneezing, like pepper." The tree has the star quality of attracting attention wherever it grows, and evokes such parochial pride that it might be called the ten-mile-limit tree ("Have a good look, stranger; you won't see another 'un like it more'n ten miles from here"), though actually it is distributed in favored locations from the Mexican border to the Columbia River. Douglas hoped that the California bay would be "a most valuable addition to the garden," but it has proved disappointing in cultivation outside its natural range.

For some days, as the brigade moved slowly southward, glimpses of the Umpqua Mountains, rising ridge on ridge in the misty distance, were caught through the trees. The mountains are unique as one of the few east-west transverse ranges on a continent ribbed by north-south systems. To negotiate across them was to clamber up and down the ridges of an enormous washboard.

As the cavalcade reached the first slope and started to climb, it began to rain. McLeod and Douglas in the lead, followed by the mounted hunters and the pack horses, found the going slippery. "The footing for the poor horses [was] very bad," Douglas writes. "Several fell and rolled on the hills and were arrested by trees, stumps, and brushwood." Naturally he wasn't subjecting Dr. McLoughlin's horse to such hazards. He not only took to his feet but "thought it prudent to carry my gleanings on my back, which were tied up in a bear's skin."

The ridges were hardly more than rolling foothills, averaging some twenty-five hundred feet in height, but conditions made climbing laborious. "Every two or three hundred yards called for a rest, and two or three times in the course of the day [a stop] for a pipe of tobacco." Much of the resting was to allow the pedestrian contingent to keep pace; for they were entering the territory of the Umpquas, a different breed of Indian from the well-disposed tribes of the lower Columbia.

Living conditions were pinched in the southern mountains. Salmon ran irregularly and often not in numbers sufficient for accumulating stores of dried fish. The hill people relied on deer and beaver for a substantial portion of their diet, and didn't take kindly to strangers who trapped and shot their food reserves without permission. Hudson's Bay parties had been attacked on more than one occasion, and it was a sensible precaution to keep the brigade concentrated and alert for danger.

After three days of climbing and descending ridges, the first objective of the expedition came in sight—the Umpqua River, a wide, fast-moving, rock-filled, unnavigable stream. On the afternoon of October 17 the brigade straggled down the last slope and pitched camp at the junction of the river and a sluggish tributary known as Elk Creek, near the site of the present town of Elkton, Oregon. A stopover of several days was planned, so Douglas accompanied the trappers up the creek in search of beaver.

They were looking for "bank beavers," animals that took advantage of torpid waters to dig chambers underground. Though the instinct to dam running waters and build lodges in the resultant lakes is strong in beavers, they can also adapt themselves to still or stagnant waters by burrowing techniques more usually associated with their gopher and woodchuck relatives. The one standard requirement for either dam or bank beaver is a submerged front door from which a series of chambers lead to higher and drier ground.

Soon the hunting party saw tangible evidence of their quarry, trees gnawed off two feet above the ground, scattered woodchips, and the telltale "drag trail" where the succulent branches had been hauled to the stream's edge. At these locations, known as "landings" or "come ashores," preparation of traps would begin. The standard procedure was to fasten the metal-jawed devices to logs sunk approximately a foot beneath the water. Above this, a fresh twig held by a cord would be suspended so that it dangled some eighteen inches above the surface. As a final refinement, the last priming of the lethal mechanism, the twig was smeared with a substance known as castoreum.

The history of man's treachery toward animals is long and inglorious. The bull moose is lured within gun range by imitating the call of a mate; tigers are decoyed to their destruction by staked-out lambs; monkeys are captured by placing food in bottles that will not permit the exit of a closed fist. But in all these ruses, man tempts the creature by the prospect of some gain, by the hope of food or sex.

Only beavers allow themselves to be totally deluded and rush toward destruction driven by an appetite for something they already possess. This is because castoreum, the sweet, musky-smelling substance used to smear the dangling twig, is cut from a beaver's body. It is a genital secretion—both males and females are endowed with the glands—and clearly, in natural use, the perfume released must add to the excitement of mating. But castoreum, extracted and offered as bait, affects the animal in a way that seems to go far beyond sexual excitation. Its efforts to reach the substance are described as "frenzied," and, in cases where the smeared stick has been captured, it is orgiastically consumed, leaving the animal in a "drunken" state.

What the trapper counts on happening, and what most often does happen, is that the beaver, smelling the castoreum, swims toward the bait. Unable to heave itself out of the water far enough to reach the twig, it instinctively thrusts its rear feet downward, searching for a platform, and finds the jaws of the trap.

The beaver responds to the sudden shock of the trap by diving. It can stay submerged for as much as fifteen minutes; but if by that time it hasn't disentangled itself, it must surface for air. At this stage the beaver will gnaw its foot off to escape, if given time to think about it. The experienced trapper knows this and seeks to prevent it by attaching a large rock or other weight to the trap. When the beaver dives, it dislodges the weight which drags trap and animal down to deep water, thus insuring death by drowning.

Beaver flesh was considered the tastiest of fare, old trappers likening its flavor to pork. The animals averaged forty pounds in weight, so when the trapping was good the brigade feasted. We must assume that beavers were hard to come by that October on the Umpqua since, during the entire stay at the encampment, McLeod and his hunters continued to scavenge for deer.

Douglas lent his sharpshooting assistance for only the first few days. He had been climbing various promontories to inspect the terrain and, observing that "the country toward the upper part of the river appears to be more raised and mountainous," decided that a side expedition up in that direction might "afford my wished-for pine." He broached his project to McLeod, who agreeably provided him with "one of his Indian hunters, a young man about eighteen years old, as a guide." Since the Indian knew the country and could speak the Umpqua language, he seemed an ideal companion; but a difficulty arose—Douglas's old difficulty of exercising command. On the first day out, needing a raft to cross a flooded stream, Douglas was unable to persuade his subordinate to wield a hatchet; as a result, he spent an hour chopping timbers himself "in the course of which my hands were in a sad condition with blisters." Unable to continue, he wrote a note to McLeod asking for a sharp axe and a willing woodsman and dispatched the Indian to deliver it.

While waiting for help to come, Douglas stalked a buck, winged it, and followed up for the kill. Running through bushes, he stumbled and fell into a deep gully. Though he was knocked unconscious, his moans were heard by passing Indians who carried him to their lodge and made him comfortable till he revived. He had probably broken at least one rib because he was aware of a pain across his chest which made movement agony. Nevertheless he insisted on returning to McLeod's camp and, using his rifle and a stick as crutches, hobbled along until he met help in the form of a hunter from the brigade, who had been dispatched with an axe to search for him.

Back at camp, Douglas rested for a day, prescribing for himself hot tea and that ultimate panacea of pre-Victorian medicine—phlebotomy ("bled myself in the left foot"). He also "bathed in the river and find myself much better."

Then he received upsetting news. McLeod's original instructions from Fort Vancouver were to follow the Umpqua River south to its source; afterward he was to swing east and north, trapping his way back through the inner ranges until he reached the Columbia again at about the Hood River. It was within this mountainous loop that Douglas expected to find the sugar pine. But now he learned to his dismay that McLeod intended to change plans; he meant to head due west for the ocean.

McLeod knew that he was risking Dr. McLoughlin's wrath in departing from orders. He had led a brigade down the sea coast the previous summer and returned with scarcely enough skins to pay for the trip. The present expedition was, in a way, an opportunity to make up for that summer's failure, and good sense suggested that he should carry out his instructions to the letter. But good sense no longer weighed with McLeod since he had learned from local Indians of a mighty, deep-flowing river that paralleled the Umpqua and emptied into the ocean below it. He couldn't rest till he'd found the estuary and pushed a canoe up its unexplored waters, perhaps into history.

Douglas had no option but to go along with the brigade, though he wasn't happy about it. "Find myself much broken broken down by the day's march," he records at the first campsite. It is an expression of mood, a depression of spirit, since he is no longer suffering from his recent injury; "the pain [is] entirely gone [leaving] only a stiffness in my shoulders." His gloom deepened as each day carried him farther from his goal. Judging from entries in his journal, he collected few plants and hardly looked at the scenery, though he was forced to take notice when the cavalcade encountered the garish tribes of the lower Umpqua. "The women are mostly all tattooed," he records, "some in lines from the ear to the mouth, some across, some spotted, and some completely blue; it is done by a sharp piece of bone and cinder from the fire."

The men were less gaudy, but they showed more hostility to the interlopers, gathering in ominous, spear-carrying groups as the brigade passed. McLeod was kept busy smoking pipes of peace, distributing gifts, and assuring the chiefs that his people were only traveling through on their way to the ocean. Grudgingly the Indians traded salmon and venison for tobacco and blue beads, but the commerce was far from friendly.

At one wayside palaver Douglas noticed a chief munching sugar pine seeds. In response to his eager questioning, he learned that the seeds came from the upper hill country, that the chief wasn't planning to return home immediately, but that he would gladly send his son to guide Douglas to the giant pines.

It was too good an opportunity to miss. McLeod was somewhat doubtful about the wisdom of traveling into unknown country with an untested guide, but Douglas was so insistent that he reluctantly gave his consent and equipped the young men for a trip back up the river.

For once an expedition engineered by Douglas progressed smoothly, probably because the chief's son took firm control from the start. He stopped off at lodges along the way to bargain for additional supplies: fish, which he cooked into sumptuous meals, and blankets to fend off the mountain chill. At one point he took a detour from the river to spend the night in what turned out to be his own village. Understandably he was concerned as to how his bride of a few weeks was making out with her mothers-in-law —fifteen of them, as his father was an important chief.

On the third day of travel they began to climb the ridges in earnest. They were in pine forests now, with only ferns peeping through a thick carpet of needles. Douglas kept hoping for a glimpse of the huge coned trees each time their horses turned a shoulder of the mountains, but each time he was thwarted.

Late in the afternoon it began to rain. The men pitched their tent but "the rain, driven by the violence of the wind, rendered it impossible . . . to keep any fire," hence they were forced to eat a cold supper of dried salmon. Later, the storm increased in fury and blew the tent down. The men crouched under the tumbled and flapping canvas, protecting themselves as best they could.

In spite of the discomfort, Douglas found literary inspiration in the turbulence of the night. He entered in his journal an account of the storm, which is by far his best descriptive passage:

> Last night was one of the most dreadful I ever witnessed . . . Sleep of course was not to be had, every ten or fifteen minutes immense trees falling producing a crash as if the earth was cleaving asunder which with the thunder, peal on peal before the echo of the former died away, and the lightning in zigzag and forked flashes, had on my mind a sensation more than I can ever give vent to.

The storm moderated and before sunrise it was "clear, but very cold." The men lit a fire and were partially thawed by ten o'clock when they saddled their mounts and set off once more. They traveled eighteen miles up and down slopes and across streams. At one point Douglas was "seized with a severe headache and pain in the stomach, with giddiness and dimness of sight." Perspiration followed, during which he seems to have sweated out a fever "and in the evening felt a little relieved."

Though Douglas dates his journal in a daily sequence from October 23 through the 26th it's likely that he wrote all four entries at one time, on the evening of the 26th when, as he describes the scene, he was "lying on the grass with my gun beside me, writing by the light of my Columbian candle–namely, a piece of wood containing rosin." The discovery of the sugar pine was behind him then and he could view in perspective and with satisfaction the dramatic circumstances leading up to and surrounding the event. He could also indulge his penchant for self-commiseration and congratulation which would hardly have been seemly if success had not crowned his efforts. "When my people in England are made acquainted with my travels," he wrote, "they may perhaps think I have told them nothing but my miseries. That may be very correct, but I now know that such objects as I am in quest of are not obtained without a share of labour, anxiety of mind, and sometimes risk of personal safety."

He was entitled to his moment of grandeur. He had shown throughout his stay on the Columbia an enthusiasm and persistence that rarely flagged in spite of weather, illness, or paralyzing attacks of loneliness. As an admirer was to put it years later, he demonstrated an "indefatigable perseverance in almost every species of danger, privation and hardship for the advancement of knowledge." On that evening of October 26, 1826, Douglas was aware that he had just made his greatest contribution, so far, to knowledge.

The actual discovery of the sugar pine was surprisingly casual, even accidental. The fateful morning dawned "dull and cloudy," but promising enough to begin drying out clothes and blankets which had been wet for so long. Douglas was eager to get on the trail again, sensing that his goal was near, but the chief's son was adamant; the horses needed a day of rest even if the men didn't. However, there was nothing to prevent Douglas, if he was bent on wasting his energies, from traveling anywhere he wanted on foot.

Somewhat miffed, Douglas moved off on a southeasterly course hoping to find a summit from which he could get a clear view of the country. The location was somewhere east of the present city of Roseburg, Oregon, a landscape of tree-covered ridges stretching like waves in every direction. After an hour's walk he met an Indian who at first seemed hostile but was won over by gifts of tobacco and beads. "With my pencil I made a rough sketch of the cone and pine I wanted . . . when he instantly pointed to the hills about fifteen or twenty miles to the south." With "much good-will" the Indian offered to act as Douglas's guide, and the two men set off.

They reached the hills about midday. Coming out into a clearing, Douglas saw sugar pines for the first time. He gazed in astonishment. He had been expecting large trees, but nothing like these.

Later, describing them for his journal, he adopted a cautious approach. "New or strange things seldom fail to make great impressions," he sagely observed, "and often at first we are liable to over-rate them." To prove this wasn't so in his own case, that his objectivity was still intact despite the stunning impact of the trees, he proceeded immediately to give dimensions, choosing a wind-felled giant to measure so there could be no question of guesswork. "Three feet from the ground, 57 feet 9 inches in circumference [enough to contain an elephant] . . . extreme length, 215 feet [twice the height of a factory chimney]."

These were facts; yet even as he scrambled around the fallen tree, measuring and taking notes, Douglas must have wondered who in the British Isles, where the Scotch pine at seventy-five feet is the tallest known, was ever going to believe him.

He had everything he needed now—dimensions, leaves, samples of the bark and gum—except for cones. He could see them high above him, hanging from the branches "like small sugar-loaves in a grocer's shop." He resorted to his old technique of collection by rifle fire and succeeded in clipping off three cones with his first shot. He was reloading when a development occurred that, as he graphically put it, "nearly brought my life to an end."

His shot had attracted company—eight Indians "painted with red earth, armed with bows, arrows, spears of bone, and flint knives," who moved ominously into the clearing. Douglas pantomimed his peaceful intentions, indicating that he was only interested in pine cones. He offered tobacco.

The Indians filled their pipes and sat down to smoke, apparently satisfied with the situation. Douglas finished loading his rifle and was about to fire again when he noticed "one [Indian] string his bow and another sharpen his flint knife."

Convinced that trouble was brewing, Douglas "went backwards six paces and cocked my gun, and then pulled from my belt one of my pistols, which I held in my left hand." This display of weaponry seemed to puzzle the Indians more than anything else. After a ten-minute staring contest "without a word passing," the leader of the group made a sign for more tobacco.

He was probably only asking payment in advance for any cones shot down. As Douglas well knew, the seeds were delicacies much relished by the Indians and, in the Umpqua country particularly, were hard to come by. But with that blindness to "native" rights that seemed to afflict most nineteenth-century travelers, he was unable to see himself as either a trespasser or poacher on tribal property. Instead, he reacted with indignation at what he considered blackmail, but also with discretion because he was outnumbered. He indicated by signs that he would give tobacco only in reward for pine cones. This proposition was perfectly acceptable to the Indians who dispersed to look for fallen fruits.

Once alone, Douglas lost no time in picking up his botanical specimens preparatory to making a hasty exit, but he had forgotten his guide of the morning who now emerged from hiding ready for further service. Suddenly alarmed by all dark faces, Douglas pressed tobacco into his hand and sent him in an opposite direction. Then he took off down the hill at a run to regain his camp, soon after nightfall.

He wasn't at all sure that he was safe yet, even with the sturdy backing of the chief's son. He stayed awake most of the night alert for an attack that he felt certain had come when "two hours before day" his companion "uttered a shriek." Douglas sprang up, rifle at the ready, "thinking . . . the Indians had found me out," only to discover that the chief's son "had been attacked by a large grizzly bear." He found the situation anticlimactical. "I signed for him to wait for day," he tells us wearily, "and perhaps I would go and kill it."

McLeod was not in camp when Douglas returned to the Umpqua River. He had traveled down to the coast, hot in pursuit of his great river, but he returned in a few days, mud-stained and saddle-weary, though in surprisingly good spirits considering the disappointment he had suffered. The Indians had misled him, he explained to Douglas, or perhaps they hadn't understood. They had taken him to explore two waterways that flowed into the ocean south of the Umpqua (probably the present Coos and Coquille rivers), but neither was more than an inflated mountain stream. Before turning back, however, he had learned of another artery still farther to the south described as wide enough for a hundred war canoes to paddle abreast. McLeod, his hopes stirred afresh, was driven by the compulsion to investigate, but he had also begun to worry about Dr. McLoughlin; hence his return to camp to send a detachment back to Fort Vancouver with a report on his activities.

Douglas decided to join this detachment. It would have been exciting to continue on with McLeod, perhaps to share with him an historic moment of discovery,* but it would also have been an indulgence. With the seeds and specimens of the sugar pine safely in his saddle bags, his work in the Umpqua country was finished. It was time to think about and prepare for his return to England in the spring. A long march across the continent and an ocean voyage still lay ahead of him.

It is doubtful that he thought any longer about the girl who had once filled his mind. He may still have felt the pull, the overwhelming excitement he had never known before and was never to know again. But if he had rethought the matter, and inevitably, given his cautious nature, he had, he must have decided that his best interests dictated staying away.

Whatever his resolutions and intentions, he was unprepared for the news that awaited him at Fort Vancouver, or the successive shocks that were to unnerve him in the weeks to come.

* This was not to be; there is no navigable river between the Columbia and the Sacramento, which empties into San Francisico Bay some six hundred miles to the south.

7

DARK LADY OF THE CHINOOKS

THOUGH TWENTY-SIX WHEN HE reached the Columbia, Douglas, at least as far as his journal and letters reveal, seemed unaware of the existence of a second sex. Incredibly, after four months on the river, he had still failed to mention Chinook girls, a lively feature of the local scene and the delight of Fort Vancouver's unattached males.

Of course much of his reticence was due to the standards of his time. Though as a naturalist he was expected to take a scientific interest in the customs and habits of primitive peoples, there was such a thing as getting too detailed. Amusing anecdotes about quaint notions and backward attitudes that could make English ladies chuckle were appreciated; matters involving vulgarity were not, and certainly Douglas had no wish to appear coarse. But there is the strong probability that Douglas chose to ignore the ever-present Chinook girls not for moral reasons but because of the uneasiness they caused him. Raised in a puritanical Scottish village, frightened by fulminations against "sin" both in kirk and at home, there is more than one indication that Douglas grew up with a deep mistrust if not actual fear of relationships with women. When Professor Hooker praised his star pupil's "singular abstemiousness," the definition might have covered his sexual shyness even more than his restraint with a bottle.

Needless to say, Douglas's attitude was far from typical of other whites on the Columbia. No timidities inhibited the junior gentlemen and *voyageurs* of Fort Vancouver any more than they had sailors off ships that had plied the river since its discovery. The men of the Lewis and Clark expedition, finally reaching the end of a continent and, seemingly, the end of the earth, were startled to note the names of former sojourners—"J. Bowman" was one—tattooed on the arms of Chinook girls who rushed forward to greet them. "Sport," as William Clark had dourly called it, became the favorite recreation of that stormy winter of 1805–6, and its popularity remained undiminished by the time Douglas arrived upon the scene some twenty years later.

During his early travels up and down the river, he must have been hard pressed at times to evade nocturnal companionship, or "all the hospitality Indian courtesy could suggest" as he was later delicately to put it. But there is no doubt that he succeeded. Psychological factors aside, he hadn't come halfway around the world to compromise his respectability and jeopardize his future by frivolous conduct.

His continence puzzled the Chinooks, amusing the men but considerably irritating the women. We know that when Meriwether Lewis refused the favors of a chieftain's sister, sent to him in payment of a debt, the girl was highly indignant at the disgrace to her and dishonor to her brother, and could only be consoled and sent away when loaded down with extravagant gifts. So now a new generation of Chinook girls must have considered Douglas's restraint a slight to their charms. Finally they pieced together a face-saving theory: as a collector of knickknacks for a mad king, he had taken an oath of celibacy as part of some purification ritual.

It was an explanation that might soothe the vanity of women, but it didn't convince Cockqua, an astute Chinook chieftain of an up-country tribe. Then in his late forties, Cockqua had observed that white men's class distinctions extended down even to their sexual behavior, lower ranks tending to be more promiscuous than their superiors. White men of the highest status, such as Dr. McLoughlin and other senior officers at Fort Vancouver, seemed to make a virtue of living in monogamy, a surprise to Indian chiefs who found both prestige and variety in multiple wives.

Cockqua well remembered the aloofness of the American leaders Lewis and Clark while their men made merry, and there had been a more recent example of a high-ranking white man's strange penchant for celibacy. The year before Douglas's arrival, George Simpson, governor of the Hudson's Bay Company, had refused a temporary bride during a visit to the Columbia, even though a war canoe and a hundred beaver skins had been offered by her wealthy parents as added inducements.

Cockqua, described by Douglas as "the principal chief of the Chenooks [Chinooks] and Chochalii * tribes," made no attempt to explain such irrational, upper-class behavior, but the fact that Douglas was adhering to the pattern led him to believe that the King's collector was a greater chief than showed on the surface, thus opening up exciting possibilities.

For Cockqua, like every other tribal leader, yearned for a white son-in-law. There was wealth as well as prestige to be gained from such a connection. The ancient and wily Concomly, a patriarch of the lower Columbia, had pioneered the way by marrying no less than two daughters to white traders; as a result he enjoyed a steady supply of gold and red uniforms, copper and brass cooking utensils, hatchets, saws, and all the tobacco he and his relatives could possibly smoke.

Other chiefs had scrambled to emulate Concomly's success, and one had even landed a Hudson's Bay Company chief factor, an epic triumph. It was time for Cockqua to make his move if he intended to maintain his position as an important tribal leader, and that spring and summer, as he studied ways and means, he had every confidence for success.

Douglas left Fort Vancouver and started downriver in mid-July, 1825. It was his second trip to the Columbia's mouth since his arrival in April, but this time, escorted as usual by his *voyageur* and two Indians, he continued on to the north.

In response to cordial and repeated invitations, he was on his way to visit Cockqua's village, but he was in no hurry to get there. He planned to turn the trip into an extended botanizing tour up the coast, perhaps reaching as far as the Hoh River which, he had heard, flowed through a fern paradise. But somewhere near the present Pacific Beach he turned back, discouraged by the weather.

"Almost continual rain," he complained. "I was unable to keep my plants and blanket dry . . . Only two nights [without rain] during my stay on the shore." Today we would not be as dismayed as Douglas by the continual downpour because we recognize that he was traveling along the outer perimeter of the Olympic peninsula, the wettest region in North America, deluged by over a hundred inches of rain annually.

He turned back and portaged across the isthmus near the present Ocean City to Grays Harbor. Then, as his canoe crossed the bay and entered the Chehalis estuary, a miracle happened—the weather cleared. The sun was warm and bright when Douglas stepped ashore at Cockqua's village. "Im-

* Also spelled "Cheecheeler," "Chickeeles" and "Tsihalis." The name is now stabilized as Chehalis, meaning "sandy," a reference to the sandy-bottomed river. The Chehalis tribal groups were offshoots of the Columbia Chinooks, identical in appearance, language, and customs.

mediately after saluting me with 'clachouie,' their word for 'friend,' or 'How are you?' and a shake of his hand," Douglas was taken to the chief's lodge where "water was brought immediately for me to wash, and a fire kindled."

To accommodate his guest, Cockqua had thoughtfully partitioned off a section of his lodge—"built for me a small cabin" as Douglas described it—where some privacy might be obtained from Indian communal life. Douglas finished his ablutions in the unexpected luxury of these quarters, hung up his wet clothes to dry before the fire and rejoined Cockqua for the next phase of his entertainment.

He was taken to the wharf to view a freshly caught sturgeon, "10 feet long, 3 at the thickest part in circumference, weighing probably from 400 to 500 lb." After Douglas had duly admired the miniature whale, Cockqua asked "what part should be cooked for me." Douglas deferred to his host for judgment of "savory mouthfuls, which he took as a great compliment." At the feast which followed, segments of the spine and head provided Douglas with "the most comfortable meal I had had for a considerable time."

It was evening now and time for the climactic entertainment. By the light of bonfires three hundred warriors in full paint and battle dress filed out and took up formations. Cockqua explained that "he was at war with the Cladsap [Satsop] tribe,* inhabitants of the opposite bank of the river, and that night expected an attack." While women and children clapped hands rhythmically, the braves "danced the war dance and sang several death songs, which to me . . . imparted an indescribable sensation." Anthropologists may yearn for more details—it was rare for a white man to see such sights—but Douglas frustrates them by ending laconically, "The description would occupy too much time."

The evening was finally over. We can imagine Douglas expressing appreciation for his entertainment and Cockqua, versed in basic English as well as "European manners," replying suitably. We can see the men enter the chief's lodge and, at the entrance to the private quarters, the host bid his guest good-night.

Then something unexpected occurred. Douglas, upset, emerged from his "small cabin" and stalked outside. He put up a makeshift tent at the edge of the village and there spent the night.

What happened?

The explanation Douglas gave in his journal was that, though Cockqua "pressed me hard to sleep in his lodge lest anything should befall me"

* Douglas here confuses tribes, just as later he confuses landmarks en route to the Chehalis River. It is difficult to avoid the suspicion that the confusions are deliberate, to obscure the exact location of Cockqua's village, which was across the river from the present town of Aberdeen, Washington.

(a reference to a possible enemy attack), he decided it might look cowardly to rest snugly under a roof when danger threatened. Accordingly, "as fear should never be shown I slept in my tent fifty yards from the village."

It was hardly the most convincing of reasons. For one thing, night attacks were unknown in tribal warfare; for another, if danger really threatened, why didn't Douglas mount guard instead of sleeping, either in lodge or tent? Three years later, back in England, Douglas apparently saw the fallacies in his original thesis and experimented with a new explanation for his flight from Cockqua's lodge.

Fleas became the culprits—"immense number[s] of fleas." Because of their hoppings and bitings and "the great inconvenience suffered thereby" he "preferred to put up at my own camp, a few yards from the village, on the shore of the river." But though fleas offered a rational reason for flight, Douglas was reluctant to give up the heroism-in-the-face-of-danger motive completely, so he combined the two themes—fleas drove him to the river's edge but, once there, he was exposed to the hazards of enemy attack. Cockqua also turns up in this second version: "He was so deeply interested in my safety that he watched himself the whole night."

The Cockqua addition is interesting. We can well believe that he followed his guest to his new encampment, but we suspect that it was more to soothe his ruffled feelings than to watch over his physical welfare. Something had happened in the "small cabin" to shock and perhaps outrage Douglas's sensibilities, but what? We must wait for clues in the events of the following day when Cockqua obviously made great efforts to restore Douglas's equanimity.

Next morning (and we're back to the original journal now) Douglas tells us that he was publicly hailed by Cockqua as "a great chief, for I was not afraid of the Cladsaps [Satsops]." Apparently Cockqua had agreed to go along with the heroic motif as a face-saving device, though the facts of what actually happened must have traveled fast and hilariously through the village.

A shooting contest was arranged: Douglas, armed with his fowling piece, against a Chinook archer. This amazing marksman "passed arrows through a small hoop of grass 6 inches in diameter, thrown in the air at a considerable height by another person." Despite this demonstration Douglas could still remark complacently "of shooting on the wing they have no idea." To show how it was done, he charged his gun with swan shot, walked within forty-five yards of "a large species of eagle . . . threw a stone to raise him, and when flying brought him down."

Desperately trying to prove himself, Douglas seems to have been at his obtuse worst that morning. He was apparently totally unaware of the suppressed mirth that surrounded him and that found vent at the slightest excuse. So, when he brought down the flying eagle, he is able to report with uncomprehending satisfaction; "This had the desired effect: many of them placed their right hands on their mouths—the token for astonishment or dread." The description so closely parallels the reaction of his canoemen to the shenanigans with the Epsom salts and spectacles that we suspect that he was being ribbed again.

Douglas, unconscious of byplay and beginning to enjoy himself, next asked a warrior to throw his hat in the air; he shot "the whole of the crown away, leaving only the brim." That brought down the house; the crowd was convulsed by cheers and laughter. "My fame was sounded through the camp," Douglas tells us. "Cockqua said 'Cladsap [Satsop] cannot shoot like you.' "

Through all these sporting heroics, Douglas is getting around to something. He might have avoided it; there were other ways to explain subsequent developments without actual lying. But he seems drawn toward the revelation as though by a compulsion to get it on record, no matter what half-truths or evasions might be necessary to protect himself later.

He approaches the matter nonchalantly. "In the lodge were some baskets," he observes, "hats made after their own fashion, cups and pouches, of very fine workmanship . . . I received from [Cockqua] an assemblage of baskets, cups, &c., and his own hat, with the promise that the maker . . . would make me some hats like the chief's hats from England."

And then, tucked away in a parenthesis, we learn that "the maker" was "a little girl twelve years of age, a relation of [Cockqua's]."

At last we are on solid ground. With the appearance of a girl, all the parts of the puzzle—the flight from Cockqua's lodge, the chief's pursuit to the river's edge, and the mirth of the villagers—begin to make sense. Even the fact that Douglas was unable to exclude mention of her from his narrative impresses us with a sense of her importance.

She was a weaver of artifacts "of very fine workmanship," which alone would imply high status in a Chinook tribe. But she was also "a relation" of a tribal chief—almost certainly Cockqua's daughter, since she lived in his lodge. In short, she was a uniquely endowed and prized princess, whose fame must have traveled far and whose hand must have been often sought in marriage.

For nothing can be more certain than that she was of marriageable age. We must not be misled by Douglas's attempts to pass her off as a precocious child; the facts of tribal customs and culture are against him. She would never have been allowed to manufacture eating utensils before her first menstrual period—or indeed after it, until she had undergone elaborate purification rites. The taboos were strict on this point. An uncleansed girl, unless carefully regulated, could cause salmon to shun a river, thus threatening starvation for her tribe. Puberty rituals called for an isolation period of five months, after which the girl was welcomed back to the village and honored with a feast celebrating her new status as a marriageable woman.

Then there is the matter of acquiring weaving skill. It often took years to master the intricate designs and patterns traditional to a tribe; as a result, accomplished artists were generally middle-aged or older. Even granting unusual nimbleness of mind and fingers to Cockqua's daughter, she must have served at least a minimal apprenticeship to be capable of the superior work Douglas vouches for. So the conclusion is clear. The "little girl" emerges from her chrysalis in a new form. She may have been twelve by some obscure method of Chinook calculation, but in European terms, Douglas's terms, she was older, and perhaps considerably older.

Though he doesn't get around to telling us about this intriguing person until the second day of his visit, Douglas was certainly aware of her earlier. Cockqua would have seen to that. Perhaps it was she who brought Douglas water and kindled a fire in his "small cabin" soon after his arrival. She would not, of course, have sat with the men at the feast where the ten-foot sturgeon was consumed, but she might have kept in evidence by heaping tidbits on Douglas's plate and, afterward, during the war dance, by bringing a blanket to protect him from the evening chill.

Douglas's shock later that night was not at being confronted by a stranger, but because of panic in a situation with which he was unable to cope. The next day, however, and in the days of continued sunshine that followed, he seemed to have overcome much of his fear and even to have edged, with whatever trepidation, toward a relationship.

He tells us nothing about all this, of course, but the clues escape. For instance, during the week that he botanized around the village, there is more than one indication that he wasn't alone as he canoed through the salt marshes dispersing wildfowl in clouds of flight. More than half the specimens he collected were reeds, rushes, and bog grasses, plants of doubtful horticultural value but indispensable to the weaver's craft. Then there was his sudden interest in a weed of swampy places known as the cow parsnip. A powder extracted from the dried roots of this plant was widely used by Chinook girls as a love charm. But whereas other philters were surreptitiously slipped into food or drink, parsnip powder for maximum effectiveness had to be sprinkled on the loved one's clothes and hence no doubt was brought to Douglas's attention. Finally we hear no more nonsense about heroic sleeping in a tent by the river. The "small cabin" seems overnight to have cleared itself of fleas.

Once back at Fort Vancouver, caught up again in the atmosphere of practical affairs, Douglas was probably more than willing to let his summer idyll slip back into memory. But Cockqua had other ideas. In early September, he journeyed down from the Chehalis River and made his way up to Fort Vancouver. Learning that Douglas had left for the Grand Rapids, he set out doggedly after him.

"Last night my Indian friend Cockqua arrived here from his tribe on the coast," Douglas entered in his journal. "[He] brought me three of the hats made on the English fashion,* which I ordered when there in July . . . I think them a good specimen of the ingenuity of the natives and particularly also being made by the little girl, twelve years old, spoken of when at the village."

He paid "one blanket (value 7s.)" for the hats and threw in "a few needles, beads, pins, and rings as a present for the little girl." But if he thought that concluded their business, he had misjudged Cockqua. The chief had come a long way and had much on his mind.

It was chilly that day, forty-three degrees Fahrenheit at noon, so Cockqua must have welcomed the dram of rum Douglas offered him. The men lit their pipes and the chief began his pitch.

What did he offer Douglas to return to his village and accept his daughter as a bride? He was rich; we have Captain William Clark's word for that. Twenty years before, when the tribes of the lower Columbia region were listed and described by Lewis and Clark, the Chehalis groups were recognized as the most wealthy and populous in the Chinook nation. So now Cockqua, aware that Governor Simpson had turned down a hundred beaver skins and a war canoe as a dowry bid, must have outdone himself in generosity. We don't know the details, but again a war canoe seems to have figured in the tender, since it turns up later in the narrative.

We hope that Douglas let him down gently. His line of refusal was obvious; unlike other white men who had contracted marriages with Indian girls, he was to be a resident of the country for only a short time. To form a connection would be unfair both to Cockqua and his daughter.

Then the chief told Douglas something that must have startled him. Fairness was no longer an issue with the girl; she had totally committed herself to Douglas, hardly sleeping or eating since he left the village. Cockqua's major concern now was for her health and sanity.

He must have told Douglas that, or something very much like that. It's the only thing that could make sense out of what was to happen.

In the end, accident took Douglas back to the Chehalis River a little

* Probably modeled on the wide-brimmed straw hats worn by sailors of the period.

more than a month after his conversation with Cockqua. He need not have gone, and certainly caution should have warned him to stay clear of a compromising situation. But it is likely that his curiosity about the girl's love and the temptation to enjoy it were too strong to resist. It seemed incredible that he, who thought himself incapable of passion, should have aroused it in another.

The accident that took him back was the organization of a brigade to investigate the beaver populations of the Chehalis and Cowlitz rivers. In charge was a young Hudson's Bay Company officer named Alexander McKenzie,* recently married to one of Chief Concomly's daughters. Under these convenient and sympathetic circumstances, Douglas evidently felt that he could return to Cockqua's village without attracting undue attention.

The expedition left Fort Vancouver in late October. As before, Douglas, with the specter of Joseph Sabine reading over his shoulder, made deliberately obscure in his journal the brigade's destination and the route it followed. He also employed other diversionary tactics, such as recounting at length anecdotes about the party's guide and host-to-be, a chief named Tha-a-muxi. It appeared that Tha-a-muxi had sworn off alcohol because "some years since he got drunk and became very quarrelsome in his village; so much so that the young men had to bind his hands and feet, which he looked on as a great affront." He seemed only to have switched his method of achieving inebriation from rum to tobacco, however, because as Douglas tells it, "So greedily would he seize the pipe and inhale any particle of smoke in the lungs, that he would regularly five or six times a day fall down in a state of stupefaction."

When he wasn't waxing humorous about Tha-a-muxi, Douglas described the weather, which had reverted to its normal state of being awful. "A most violent hurricane set in from the west," he tells us. "The wind was so high, with heavy rain, that scarcely any fire could be made . . . Two days without food [and] my blanket being drenched in wet . . . I deemed it prudent not to lie down to sleep."

Through all this, the brigade moved steadily north, portaging from the Columbia's mouth to Willapa Bay, then portaging again to Grays Harbor. "A little before dusk" one evening it reached Tha-a-muxi's village

* No relation to the great Sir Alexander Mackenzie of exploring fame or to other McKenzies (or Mackenzies) in the Company. The duplication of Scottish surnames—McDonald (but also M'Donald and MacDonald), McLeods, McDougals, McMillans, and so forth—was so confusing to the *voyageurs* that they took to identification by nickname; thus, McDonald *le grand* (fat), McDonald *le prête* (priest), McDonald *le bras croche* (crooked armed) and McKenzie *le rouge* (red headed), McKenzie *le borgne* (one eyed), even McKenzie *le picoté* (let's say, the picador).

which, with Cockqua's and a dozen others, was located at the estuary of the Chehalis River.*

Though, for the record, Douglas claims that he "made a stay of several days at [Tha-a-muxi's] house," the facts suggest that, as soon as etiquette permitted, he slipped away to Cockqua's lodge. Certainly his satisfied testimonial that he enjoyed "every kindness and all the hospitality Indian courtesy could suggest" could only emanate appropriately from Cockqua's "small cabin." And if further evidence is needed, Douglas's list of seeds collected during those October days should serve to pinpoint where he was and with whom. For, without exception, they were seeds from plants he had identified in July, though the list is far from complete; he didn't revisit *every* plant. We suspect that, for once, Douglas was botanizing with something less than full attention. He had other things to occupy his mind.

Finally there is the evidence of the canoe. When the brigade proceeded on its way upriver, Douglas traveled in some style, in a "canoe too large to pass in many places" and which finally had to be sent back because "of cascades and shallowness of the water."

Was this a ceremonial war canoe, hollowed from a single tree and capable of accommodating up to twenty warriors? We don't know; but at least it seems to suggest that Cockqua was sending Douglas off with some of the pomp and circumstance befitting the son-in-law of an important chief.

Despite his best precautions, word of Douglas's romantic interlude inevitably filtered back to the gentlemen's mess at Fort Columbia. Probably not through brigade leader Alexander McKenzie, who seems to have acted with discretion and sympathy throughout. It is more likely that the *voyageurs* on the expedition, amazed at the breach in Douglas's formidable rectitude, had snickered and told.

Possibly, too, Cockqua journeyed down to Fort Vancouver, determined to avail himself of the perquisites of his new relationship and, by the time he had been ushered back into his canoe, loaded down with machine-made gifts and novelties, eyebrows had been raised at the officers' table. However it was, by late December Douglas was heartily sick of fruity chuckles and innuendoes, and also, of course, in panic at the possibility that news of his escapade might somehow get back to Joseph Sabine and the Horticultural Society.

* John Work, a Hudson's Bay Company chief trader who was in the area the previous year, noted fifteen villages at the river's mouth, supported by the phenomenal salmon runs.

His gloomy frame of mind is best revealed by the entry in his journal for New Year's Day, 1826:

> Sunday, January 1st.—Commencing a year in such a far removed corner of the earth, where I am nearly destitute of civilized society, there is some scope for reflection. In 1824, I was on the Atlantic on my way to England; 1825, between the island of Juan Fernandez and the Galapagos in the Pacific; I am now here, and God only knows where I may be the next. In all probability, if a change does not take place, I will shortly be consigned to the tomb. I can die satisfied with myself. I never have given cause for remonstrance or pain to an individual on earth. I am in my twenty-seventh year.

Apart from the self-commiseration, and the anger at the gentlemen of Fort Vancouver (he excludes them from "civilized society" because they twit him), the passage is interesting because of its defense against a possible charge of seducing the innocent. His conscience is perfectly clear on that point, he tells us. In fact, if there was any innocence taken advantage of, it was his.

It was hardly a gallant stand, but by this time Douglas was badly frightened. Feelings of tenderness and affection which may have developed during romantic hours were lost in the discomfort of his situation at the fort and his anxiety about his career. No doubt he solemnly promised himself never to approach within fifty miles of Cockqua's village again, or think of the girl whose name he was always to keep from us.

Yet he was to return twice more, though probably only to see her alive once.

As we have seen, 1826 was a busy and productive year for Douglas. During the spring and summer, he botanized through the Blue Mountains and up the Columbia as far as Kettle Falls. In the autumn, he accompanied Alexander Roderick McLeod into the Umpqua country, found the sugar pine, and returned through endless rain to Fort Vancouver late in November.

Despite his excellent physical condition, the constant exposure to wind and weather had taken its toll. "Swelled ankles . . . accompanied by an acute pain . . . obliged [him] to remain indoors for nine days." Then, in early December, just when he was beginning to move around again, he was summoned to Cockqua's village.

It is the only interpretation to put on it. The timing is about right; nine days would permit news of Douglas's return to the fort to reach Cockqua, and allow a vital message to travel back.

Douglas was hard put to concoct an excuse for a botanizing expedition in winter, but Dr. McLoughlin seemed to take no exception to his plan to visit the ocean "in quest of Fuci [lichens], shells, or anything that might present itself to my view," and assigned the usual *voyageur* and two Indians to him. Through "rain falling in torrents," Douglas canoed and trudged his way to the Chehalis River.

His "old Indian friend, Cockqua . . . greeted [him] with that hospitality for which he is justly noted" and gave him the usual winter fare of "dry salmon and berries" to eat. But Douglas gives no account of what happened next, nor drops hints as to the real reason for his precipitous visit.

We are forced to surmise. It had been more than a year since the happy occasion of "every kindness" and "all hospitality," and he could only have been induced to return by some powerful and unanticipated emotional pull. Also it seems certain that disputes broke out during the six days of Douglas's stay at the lodge. Perhaps Cockqua, in light of a new development, wanted Douglas to acknowledge his union as openly as Alexander McKenzie and other white men who had married chieftains' daughters.

If such was the case, Douglas unquestionably refused. Having ascertained whatever brought him to the village (perhaps the fact of a child and its possible parentage) he was now anxious to depart. Providentially, he "was seized with a most violent diarrhoea." He became "alarmed lest it prove dysentery . . . therefore . . . set off."

He reached Fort Vancouver on the tide five days later, still somewhat ruffled by his confrontation with Cockqua, but with his stomach troubles apparently cured. He entered in his journal a seeming valediction to the Chehalis River, condemning it as an area in which he had "gleaned, like my trip in the same quarter last year, less than any journey I have had in the country."

But his farewell was not to be as final as he wished. Within two months he was back at Cockqua's lodge.

It was a lightning-fast visit. Previous expeditions from Fort Vancouver to the Chehalis estuary had taken from five to nine days; this time Douglas accomplished the round trip—there and back—in eight days. Of necessity his business at the village must have been concentrated into a few hours, a span so short as to suggest a fruitless mission.

Again the trip seems to have been wholly unanticipated. In January, 1827, Douglas's thoughts were turned toward England; he was packing his specimens, clearing the way for his journey across the continent with the Hudson's Bay eastbound express. It had been snowing heavily. "The forest presents a most dismal appearance, the immense pines loaded with snow and their wide, spreading branches breaking under their load." There was more snow in February, "15 inches deep, which lay until the 25th, and after that, frequent rains and gusts of wind."

It was no weather to be outdoors if it could be helped; yet suddenly, on March 2, Douglas is Chehalis bound. He searches desperately for a reason to justify the trip, finally coming up in the journal with "my object was to procure the little animal which forms their robe." What little animal? What robe? He's too distraught to tell us.

Douglas's companion for the trip was a Hudson's Bay Company clerk named Edward Ermatinger, a sensitive and accomplished musician, and it is probable that the two men traveled alone. *Voyageurs* and Indians would hardly have tolerated the forced marches under dreadful conditions necessitated by their rate of speed.

What drove them? What had Douglas learned to launch him on another exhausting expedition to Cockqua's village just days before his departure for England? Why did Ermatinger, when Douglas confided in him, immediately offer his support and companionship?

Again we must deal in probabilities. Though Douglas tells us little, we have the testimony of his contemporaries who experienced or observed situations comparable to his. We know, for instance, that in the inevitable clashes between white and red customs and values, the outcome was sometimes tragic; cases were known where Chinook mothers, forbidden by white fathers to flatten their infants' heads, killed their offspring rather than subject them to lifelong disgrace.

We have no reason to believe that Douglas laid down any such strictures during his visit in December. On the contrary, he was probably so little concerned with the future welfare of either mother or child that his indifference was palpable and wounding. And it could have been that rejection, that denial of the claims and obligations of love, which finally snuffed out hope and a last flickering purpose for life.

Douglas had come to find "the little animal which forms their robe." He admitted failure to his journal "in consequence of one of the principal men of the village, a cousin of my Indian friend Cockqua, dying the night before I arrived." It was probably as close as he could come to a truth which, then as later, he was unprepared to face.

Chinook dead were placed in canoes which were propped on the river's bank in preparation for departure to the spirit world. Possibly that is where Douglas last saw his "little girl." She would be dressed in ceremonial robes, her baby in her arms, waiting placidly for a high tide to carry her out to sea.

8

EXILE ON HIS NATIVE HEATH

DOUGLAS RETURNED TO LONDON IN OCtober, 1827, to find that in scientific circles—the only circles he cared about—he was famous. For almost two years, botanical journals and horticultural magazines had been describing the novelties he had sent back from the Columbia, while gardeners at Chiswick and on half the great estates of England had been growing his seeds and bulbs. It took imagination to envision stately trees in still spindly seedlings, but his shrubs were maturing and many of his herbaceous introductions had produced blossoms of astonishing beauty. Of a single plant, the red-flowering currant, it could be written in a leading periodical, "of such importance do we consider it to the embellishment of our gardens, that if . . . Mr. Douglas's voyage had been attended with no other result than the introduction of this species, there would have been no ground for dissatisfaction."

Honors followed one another: he was elected to membership in the Linnean Society, the leading botanical organization in Europe; he was also voted into the Zoological and Geological societies in recognition for the work he'd done on their behalf. In all three cases initiation and membership fees were waived in perpetuity "in consequence of [his] great services . . . to Natural History." His employers, the Horticultural Society, arranged for the publication of his journal and made him a gift of all revenues that might result from the sale of the book. The council also instructed Secretary Jo-

RIBES SANGUINEUM (*red-flowering currant*)

The beautiful shrub that caused a sensation when it first blossomed in England from seeds sent from the Columbia River. It is still considered by many to be Douglas's single most spectacular contribution to gardens of the temperate zone.

seph Sabine and Assistant Secretary John Lindley to aid Douglas in any way he might require to prepare his journal for the press.

It was an auspicious beginning and Douglas was pleased at the attention shown him, though he found that the limelight could also be frightening. About a month after his return, he was asked to prepare a paper on the sugar pine for the Linnean Society. At that period, papers were normally read to the Society by a specially appointed elocutionist, but such was the interest in the celebrated collector that Douglas was apparently asked to read his own composition. On the day of the meeting, however, he failed to appear and Joseph Sabine read his paper for him.

Exhaustion was given as the reason. Douglas had recently completed a three-thousand-mile march across Canada, climaxed by a near-fatal accident in Hudson's Bay where, caught by a gale in an open boat, he had been buffeted and drenched for three days and two nights before being rescued. As an aftermath, his old enemy arthritis returned to plague him and he had been laid low in his cabin for most of the month-long crossing of the Atlantic.

But despite these strains, we find the exhaustion theory unsatisfactory. Douglas had always before shown such resilience in bouncing back from physical hardships that it is difficult to picture him as still down and out a month after his return to England with all the attending stimulation, and we are tempted to look for another explanation for his failure to appear before the Linneans.

It could be that, surrounded once again by upper-class English voices and attitudes, Douglas reverted to the state of psychological paralysis they always seemed to inflict upon him. Gone was the self-assurance with which he had, in the main, conducted himself in New York, Fort Vancouver, and other outposts of culture in the New World. Back home he was overwhelmed by misgivings and feelings of inferiority well understood by that Irish outsider Edmund Burke, who had observed long ago: "Persons . . . who have not been used to approach men in power, are commonly struck with an awe which takes away the free use of their faculties."

Unquestionably Douglas yearned to read that paper before the Linnean Society. He had earned the right, laboring long to find the sugar pine and studying hard to describe it in correct botanical language. But in the end, the prospect of standing before a room full of wellborn gentlemen presided over by a nobleman wearing a three-cornered hat and addressing them in the burred and glottal-stopped accents of the Highlands was too much for him, and he panicked.

He was to demonstrate the same ambivalence toward social opportunities. Before he left the Columbia, Joseph Sabine had written him: "You will find us on your arrival not only delighted to see you but ready to make every acknowledgement to you for your exertions and for your pains and labour." Now, in fulfillment of that pledge, the secretary was prepared to introduce Douglas around London to people of influence and importance.

We can imagine Douglas's renewed misgivings; but no doubt he hoped that, on a personal basis, with his reputation speaking for him, he could more successfully converse with his betters than from a lecture platform. At any rate, he accepted the secretary's offer and throughout the winter accompanied him on his social rounds.

The first and most important drawing room to be visited belonged to the Henry Brownes. Mrs. Browne was Sabine's sister. With her own way to make, as with the rest of that extensive brood, she had married brilliantly. Henry Browne, a Fellow of the Royal Society, dabbled in astronomy, but was better known for his social gifts and wealth. The Brownes' salon at 2 Portland Place was widely celebrated as a mecca for politicians, men of affairs, and members of the scientific establishment.

On a typical evening, the formidable Captain Henry Kater, calculator of the length of a pendulum's vibration, might be found chatting to Giddy Gilbert, the tireless M.P. who had the reputation of serving on every committee in Parliament. Or Sir Humphrey Davy, the inventor of the miner's safety lamp, might discuss electromagnetism with Astronomer Royal John Pond, though their topic was more likely to be fishing, about which both were enthusiasts. Sir John Barrow was seen at the receptions, often shouting into the ear trumpet of Edward Troughton, England's foremost designer of surveying instruments. Sir John, permanent undersecretary of the navy, a job pioneered by Samuel Pepys a hundred and fifty years before, was vitally interested in such instruments because of his responsibility for ships exploring the Arctic, where Point Barrow and Barrow Strait still memorialize his name.

Not to be forgotten were the wives, sisters, and daughters of these outstanding men, many of whom, though deprived by their times of opportunities for higher education, were noted for their intellectual capacities.

And how did Douglas fare in this brilliant company? We have no eyewitness reports, but we have two subsequently written accounts by contemporaries informing us how he conducted himself, or is thought to have conducted himself.

The first casts him in the role of lively entertainer, regaling his audiences with tales such as that about his "botanical horse," an animal which he trained "to stop when passing under branches of trees which might tear off the bundles of plants which it carried on its back." More thrilling adventures involved "running for his life from Indians, eating the skins of animals, twice even having to eat up his horse."

This chronicle was written by George Bentham, "a shy and retiring" young botanist who had probably, in his own life, never faced a situation more hazardous than a chance encounter with a bumble bee. He hardly knew Douglas, so the best that can be said about his version is that it demonstrates a lively, somewhat malicious imagination, but conveys no information worth taking seriously.

The second account paints a very different picture of Douglas in society. Without effort on his part, solely because of his fame, "his company was . . . courted" wherever he went. It was an enviable position, but the adulation took its toll. "Unfortunately for his peace of mind he could not withstand the temptation (so natural to the human heart) of appearing as one of the *Lions* among the learned and scientific men in London; to many of whom he was introduced by his friend and patron, Mr. Sabine."

This record was made by William Beattie Booth, a boy from Scone who had known Douglas all his life. In 1828, Booth was working as a clerk in the Horticultural Society's garden at Chiswick, a position Douglas may well have helped him obtain. Though admiration and sympathy for his friend shine through the sentences, we must remember that Booth, because of his class position, could have had no clear idea of the society in which Douglas was attempting to mingle. *He* may have believed, when he saw Douglas leave the gardeners' lodge in his grand clothes, that "the learned and scientific men in London" could lionize his friend, but it is more difficult for us. Booth's testimony, in short, though better intentioned than Bentham's, is no more reliable.

Under the circumstances we seem justified in creating our own image of David Douglas at 2 Portland Place and at the other houses to which Sabine may have taken him. We can see him standing stiffly at the side of a room, clad in a high collared cutaway coat with narrow trousers strapped under his insteps, a fashion established by Beau Brummell ten years earlier and not to change until Victoria ascended the throne ten years later. Secretary Sabine is at his side and, as Colonel William Mudge, the geographer, passes, he introduces the men with some quip to the effect that they have the world in common. It is doubtful that the colonel is detained long. After coolly surveying Douglas, placing him in education and class, he concludes that the young man can hardly contribute to the theoretical controversies of the day, in which alone the colonel is interested; so, after murmuring a few pleasantries, he moves off.

How often were such incidents repeated in one form or another before Douglas became discouraged? We know that, once started on the round of "evenings" and receptions under the sponsorship of Joseph Sabine, he was dazzled by the minds and style of the people who moved about him, and yearned for their acceptance. He was never reconciled to what inevitably happened. Years later he wrote with such bitterness about "my London fashionable friends" that it is apparent how deeply he was hurt.

We wonder if there was a *coup de grace*, a final shock to his ego that turned him from salon doors. Perhaps he took a politely insincere invitation to call too literally and presented himself at some drawing room unaccompanied by Secretary Sabine. He could have learned much and swiftly in the embarrassments that followed.

However it was, we cannot agree with his friend William Beattie Booth's moralistic conclusion that "when the novelty of his situation had subsided, he began to perceive that he had been pursuing a shadow instead of a reality." The chances are that Douglas still wanted that shadow and continued after it, though, in the spring of 1828, he changed tactics and set about pursuing it in a different way.

There was still his journal. He was convinced that its publication would bring new fame to revive his dwindling vogue.

John Murray of Albemarle Street was the London publisher who had undertaken to bring out Douglas's journal. It was an unprecedented honor. Murray specialized in books of travel, adventure, and exploration,* but he had never before considered a work by an ordinary botanical collector. Neither had any other publisher. Francis Masson, a collector for Kew Gardens who traveled for twenty-three years through Africa and its offshore islands, had never been approached for publication; nor James McRae, who explored Brazil, Hawaii, Chile, and Peru for the Horticultural Society.

As a matter of fact, their employers, who claimed ownership of everything their travelers wrote as well as collected, would have taken a dim view of any such disposal of rights. George Don, a collector for the Society in the West Indies, found this out when he submitted an article to the *Edinburgh Philosophical Journal* and was summarily dismissed. So, in Douglas's case, the action of the council not only in arranging publication for his journal but in permitting him to keep any and all revenues derived therefrom, was both a tribute to its regard for him and without parallel in generosity.

With his neatly filled notebooks before him, and Joseph Sabine and John Lindley standing by to provide editorial assistance as needed, it seemed reasonable to suppose that the manuscript, lengthily but fashionably titled

* For example, a two-volume work by Douglas's special hero, Captain John Franklin, titled: *Narrative of a Journey to the Shore of the Polar Sea, in the Years 1819, 1820, 1821, 1822.*

Journal of an Expedition to North-West America: Being the Second Journey Undertaken by David Douglas on Behalf of the Horticultural Society, would be ready for the presses within a few weeks. In actual fact the book remained unpublished for more than eighty years. It was not until 1914 that officials of the Society, then the *Royal* Horticultural Society, ran across the faded notebooks and manuscripts of 1828 in an old packing case and dutifully undertook publication of a work which was by that time of little more than antiquarian interest.

What had happened?

In essence, Douglas had developed a scheme for his book that was far beyond his powers of execution. It was a scheme, in fact, more ambitious than either his publisher or editorial assistants had ever contemplated. They presumed he would confine himself to the narrative portions of his journal, the day-by-day happenings interlaced with descriptions of scenery and amusing or adventurous anecdotes about animals and Indians. Botanical matters they expected him to leave for experts to ponder and publish at a later date.

They presumed all this without knowing what was passing in Douglas's mind, unaware that he had no intention of writing a popular, nontechnical book. He had a model text he hoped to follow, Thomas Nuttall's *A Journal of Travels into the Arkansa Territory, during the year 1819.* He had taken that book with him to the Columbia and knew the contents practically by heart. He admired the botanical authority with which Nuttall could roll out such passages as "I found abundance of the *Dracocephalum cordifolium* with long slolons like ground ivy, also *Hesperis pinnatifida*, but I was more particularly gratified in finding the *Tilia heterophylla*," or "On the river lands I here first noticed the occurrence of *Brunichia, Quercus lyrata* and *Carya aquata* (*Juglans*, MICH.)." Douglas wanted to fill his journal with similar passages but he lacked the technical knowledge to classify most of the plants in his collections. So at some time during his discussions with Sabine and Lindley he must have taken a deep breath and proposed that they perform the classification work for him.

The secretaries were appalled. The project involved some thousand plant species. Even if they dropped everything else and devoted themselves exclusively to the task, it would take most of a year to complete it. They tried to persuade Douglas of the enormity of his request. They pointed out that specialists like Aylmer Bourke Lambert and Professor Hooker, whose lifework it was to classify and systematize, had staffs of botanists and assistants to help them. Surely it was better to trust Douglas's rich collections to such devoted hands rather than to the part-time employment of theirs. But what they failed to point out, or probably chose deliberately to ignore, was at the heart of Douglas's reluctance to relinquish his specimens. He was always jealous of the right of others to name "his" plants, and even the fact that both Lambert and Dr. Hooker would be honor bound to name at least some of the new varieties after him didn't reconcile him to loss of total control in the matter. So in the end he decided to forego the help that Sabine and Lindley were unable or unwilling to give, and to learn enough technical botany to classify his own materials.

To his credit it must be said that, for a while, he applied himself assiduously. He concentrated first on conifers, describing some of his new pines with professional flourish. He moved on to the one new oak *(Quercus Garryana)* he had found on the Columbia, then "worked up" various other plants, seeming to choose them at random.

Unskilled at shortcuts and untrained in selective techniques, he found classification slow and exacting work. Finally, he seemed to realize the impossibility of the task he had set himself and abandoned his efforts.

He didn't admit this to Sabine and Lindley, nor did he turn over his specimens to Lambert and Hooker. He simply stopped work on his journal, both the narrative and botanical aspects of it. He may have kept hoping that somehow, somewhere, there was a solution to his dilemma. If so, he never found it.

As the summer progressed, Douglas became increasingly bored and restless. For a while, after he abandoned his literary efforts, he found a focus for his energies in propagating the seeds he had carried back across Canada, but in time that, too, palled. Discovering and collecting the seeds had won him fame; planting and tending them was gardener's work.

The presence of John Lindley, also lodged at Chiswick, seemed a continuous irritation. The esteem which had led him to commemorate the assistant secretary by several choice species on the Columbia River, an esteem enhanced by gratitude when Lindley named a genus of mountain primrose *Douglasia* in his honor, evaporated in the aftermath of their differences over his journal. Douglas appeared to blame Lindley almost entirely for the refusal to supply botanical expertise, possibly because he found it impolitic to hold a grudge against Secretary Sabine. When he learned of Lindley's appointment as the first professor of botany at London University, jealousy was added to resentment. He attended a banquet at which Lindley made an impromptu speech and promptly reported to Professor Hooker, "The beginning was bad, the end was bad and the middle worthy of the beginning and the end, not one sentence worth repeating, and the manner of delivery shockingly ill."

His peevishness surfaced again in the matter of the Society's hall porter. When Douglas learned that this top-hatted dignitary, whose main function was to relieve Fellows of their umbrellas at the Lower Regent Street clubrooms, received higher wages than were doled out to him, he raged for days.

Actually he had no cause for complaint. While he was overseas, he had been paid at the rate of £100 per annum (approximately $2,500 in modern purchasing power), a tidy enough, tax-free stipend for a young bachelor whose room, board, and traveling expenses cost him nothing. When he returned to England, the council, uncertain as to his status, retained him in his former rank and pay of gardener. It never viewed Douglas, as he apparently saw himself, as a celebrated plant collector awaiting reassignment; the council had, in fact, no plans for further expeditions in the foreseeable future.

In protest against what he considered the insultingly low pay, Douglas left the gardeners' lodge at Chiswick and moved across the intervening fields to St. John's Wood, then still a rural area, where he boarded with his sister and brother-in-law, the Atkinsons. William Atkinson, a prominent architect of the day, had been employed for decades remodeling the great mansion at Scone, where one of his apprenticing sons had met and married Douglas's sister, later bringing her up to London to establish a home. Douglas benefited from the change in atmosphere, but he still bristled enough to shoot out occasional barbs, such as to the secretary of the Zoological Society. He protested to that worthy gentleman against his name's appearing on the membership list preceded by a "Mr." and further objected to being addressed through the mails as an "Esquire." He was *not* landed gentry, he announced irritably; he wasn't even a householder, and "as it is doubtful if I eve[r] can afford a residence," requested that the secretary "give me no address" in future lists.

The truth is that Douglas, after less than a year back in England, had lost purpose and direction and was thrashing around aimlessly not knowing what to do with himself. He wanted to be sent out on another botanical mission but, as we have seen, the chances for that were slim. In a sense, he had been too successful on his previous expedition. Gardeners at Chiswick and all over Europe were straining to catch up with the wealth of material he had sent back from the Columbia and, for once, there was no demand for new exotics.

Previous collectors, after successful tours and while their reputations were still fresh, had moved on to secure and respectable jobs. John Goldie, a traveler in the United States and Canada from 1817 to 1819, had received an appointment as head gardener to the Czar of Russia and was now in St. Petersburg. James McRae, who had worked through South America while Douglas was on the Columbia, was now curator of the Ceylon Botanic

Garden. John Damper Parks, the Society's man in China from 1823 to 1824, was horticultural superintendent for the Earl of Arran.

There is no question but that Douglas, more famous than any of them, could easily have obtained a similar appointment if he had wished. There is no doubt, either that he didn't. He persisted in thinking of himself as a temporarily unemployed plant explorer only waiting for an opportunity to add new luster to his name.

Meanwhile inactivity palled. He continued to hang around Chiswick, often unshaven and sloppily dressed—"quite a *sauvage* in appearance and manners," according to a contemporary report—a burden to himself and an embarrassment to the Society, which dared not discharge its greatest traveler for fear of criticism, yet could find no satisfactory way to employ him.

Douglas's perverse mood seems to have been particularly strong when Thomas Nuttall, on one of his periodic trips to England, visited Chiswick. It should have been a gratifying occasion for the younger man. Five years before, in Philadelphia, an untried neophyte, he had been flattered merely to appear in the company of the distinguished botanist. Now Nuttall was seeking *him* out to learn at first hand about his wonderful plants from the Columbia.

It is probable that Nuttall, restrained by nature, rarely talkative, didn't enthuse openly about Douglas's collections, but there can be no doubt about his intense interest. For years the Pacific coast had been his own promised land. He had tried twice to reach it, overland through the Rockies, to be driven back each time by hunger, sickness, and the hostility of Indians. Now, at Chiswick, he was getting a preview of what he was to see some day, and the experience must have been all-absorbing. In particular, he must have been excited by the different species of *Mahonias* and *Collinsias*, genera which he had originally described and named.*

The satisfactions of Nuttall's visit should have gone far to compensate Douglas for the pangs of belittlement and neglect he had suffered in recent months, but apparently they didn't. Instead of enjoying the change of roles, which cast him as the authority and Nuttall as the attentive listener, he seems to have been oppressed by some undefined sense of injury.

It took him time to clarify his feelings, but, after ruminating on the matter for several years, he came to the conclusion that Nuttall had been

* Named in honor of his botanical friends Bernard McMahon and Zaccheus Collins of Philadelphia. It seems a shame that neither magnificient plant could have been named for Thomas Jefferson, who did so much for botany, but Nuttall harbored a grudge against the third President because of his failure to finance a botanical investigation of the Louisiana Purchase territory. The genus which was eventually named *Jeffersonia* is an insignificant plant with the unfortunate common name of rheumatism root.

too quiet during his visit to Chiswick, not nearly enthusiastic enough. Conceit, selfishness, and perhaps jealousy had prevented him from lavishing on Douglas's plants the praises which were their due. "Nuttall is a poor fellow," he summed it up in a letter to Hooker. "Next to an untruth there is nothing in the world I detest more than the narrowmindedness we too often see even in men who ought to be above such thoughts. A man of science who labours not for *self* but from an honest endeavour to add to the stock of knowledge, must of all things feel an inexpressible delight in beholding his fellowman engaged in the same laudable undertaking."

Nuttall, of course, never learned about this attack. If he had, he might have been surprised at its source—he had every reason to believe that Douglas owed him at least goodwill—but not shocked. For he had written years before, in words serving as a sad but apt comment on Douglas's assault, "I have found what, indeed, I might have reason to expect from human nature; often, instead of gratitude, detraction and envy."

Before the year 1828 ended, Douglas's prospects began to look up. Actually, the turn for the better had started late in summer when Captain Edward Sabine, Secretary Sabine's younger brother, showed up at Chiswick to carry out some scientific experiments, and Douglas volunteered to help him.

Edward Sabine, then forty, had matured during the Regency and could assume with withering suddenness the haughty style and insolent manner that characterized the fashionable buck * of the period. To his friends, however—and Douglas rapidly became a close one—he could be infinitely charming. Behind his vagaries and studied mannerisms lay a brilliant mind that had already carried him far. As a young artillery officer in Canada during the War of 1812, he had been decorated and promoted as much for ingenuity as for bravery in the assault on Fort Erie. After hostilities, he had served as astronomer to Captain William Edward Parry, another entrant in the ever popular contest to find a Northwest Passage, and had returned from the Arctic with sufficient scientific reputation to win him membership in the Royal Society. Like all the Sabines, Edward had his own way to make financially, a problem he'd solved with a minimum of wasted motion by marrying an heiress.

During the late summer of 1828, the captain was studying the dip of the magnetic needle, the puzzling variation of the compass between true and magnetic North. Douglas, assisting him, became fascinated by the intricate instruments and techniques used in the experiments, and expressed his regrets

* Also called a beau, blood, swell, and dandy.

that he had known nothing about them while he was on the Columbia, otherwise he might have been of service to the government by establishing the location of promontories and landmarks throughout his travels. Sabine was sufficiently impressed by the thought of lost opportunities to pass the information along to W. R. Hay, permanent undersecretary of the Colonial Office, in whose mind it lodged like a burr.

Douglas couldn't have spoken up at a better time. The boundary question between the United States and Canada had reached one of its intermittent pressure points. The Treaty of Ghent (1818), drawn up to settle the unresolved issues of the War of 1812, had established the border between the nations at the forty-ninth parallel, but only as far as the Continental Divide atop the Rocky Mountains. It had been agreed that, from there on west to the Pacific, the country was still too empty * for claims to be recognized; so, for a period of ten years, the territory had been left "open" for all comers.

The ten-year period elapsed in 1828 and negotiations were in progress on the thorny question. The Americans had proposed extending the boundary along the forty-ninth parallel all the way to the Pacific Ocean. The British had counterproposed a border which wobbled south down the Continental Divide to about level with the present town of Lewiston, Idaho, then made its way in a series of loops and curves down the Snake and Columbia rivers to the sea.

In point of fact, the British arguments for their proposed border, which would have gathered under his Majesty's sovereignty the present state of Washington and chunks of Idaho and Montana, were excellent. The Hudson's Bay Company effectively controlled the area, providing the only law and order under white authority. As against that great trading organization's chain of forts, manned by career personnel, the Americans could claim only a score or so of footloose trappers who had no intention of settling permanently in the country.

The British, in fact, had most things going for them except semantics. Historians have been puzzled by the tenacity with which they clung to the word "Columbia" to describe both river and surrounding territory when they had a perfectly acceptable alternative in "Oregon." The name Columbia should have been a constant, galling reminder that an American had originally discovered and taken possession of the river; but they persisted, even when forced north to the present boundary. True, they wound up with half the river, but the early, immature, tumbling half, not the great navigable artery of their expectations. Yet habit was so ingrained that they still called

* Of course it was full of Indians, but nobody bothered about them.

the province through which this minor tributary flowed British Columbia. Better to have called it Vancouver and banished all Yankee connotations.

Oddly enough, Thomas Jefferson, with everything to gain by calling it the Columbia territory, invariably referred to the area as Oregon, a name with no psychological or historic overtones. It is so noncommittal, in fact, that nobody is even sure of its derivation.

One farfetched theory that, implausibly, rings true is that the word is based on a corrupt spelling of Wisconsin. The French, having no "w" in their alphabet, spelled the word "Ouisconsin," and a French cartographer, running out of space on a map he was preparing, printed the name of the Wisconsin River in two tiers:

-sin
Ouiscon

Transposition errors and a dropped syllable accounted for the name's being copied as "Ouricon," later developing into "Ouregon." William Cullen Bryant finally stabilized the spelling when he described in his "Thanatopsis,"

> . . . the continuous woods
> Where rolls the Oregon, and hears no sound,
> Save his own dashings.

In November, 1828, Undersecretary W. R. Hay of the Colonial Office was laboring to accumulate enough facts and opinions to support the British position so that the boundary conference, which had been dragging on interminably, might be brought to a successful conclusion. He immediately saw the possibilities when Captain Sabine reminded him that David Douglas had recently been a traveler in the disputed area. Douglas, after all, was a man of science, known on both sides of the Atlantic, and his views on geographic questions should carry a certain nonpartisan weight.

Accordingly, Mr. Hay wrote the collector asking for his opinions on a "natural" boundary between the United States and Canada, meaning a boundary comprised of mountains, lakes and rivers, *not* the rigid parallels of latitude so dear to American hearts. In due course he received back a gratifying memorandum, accompanied by a letter that read in part: "There is not any natural Boundary which could give a plea to the American Government to claim this fine country up to 49°. Neither have they priority of discovery either on the Coast or in the Interior. The Boundary Line ought to extend (from my observation on the spot) from . . ."

In the remainder of his letter and, more learnedly, in the memorandum, Douglas spelled out in great detail his proposed boundary. By not quite a coincidence (he had consulted with Hudson's Bay Company directors in London) it followed almost exactly the officially sponsored line, the same corkscrew border, twisting and turning down the Rockies, then bending out to sea down the center of the Columbia, "leaving the river open to both powers."

It is doubtful if the American negotiators were much impressed by Douglas's views—certainly not enough to change their tactics, which were to stall until the conference broke up in frustration. For they knew that history was on their side; that, sooner rather than later, a land-hungry population would pour across the Rockies, making questions of occupancy academic. In 1846, when the boundary question was finally settled, the British, frightened by the American clamor for "Fifty-four-forty-or-fight!," counted themselves lucky to escape at the forty-ninth parallel.

Douglas, cheered by his association with Edward Sabine and flattered by the interest of the Colonial Office, took a new lease on life during the winter and spring of 1828–29. He found a number of stimulating activities to occupy his time—some, oddly enough, of a literary nature. Unable to make headway with his own book, he volunteered editorial assistance to more successful authors. Dr. John Richardson, naturalist of the Franklin Expeditions, was preparing a work on the zoology of the Arctic regions, and Douglas helpfully chipped in with his notes and specimens from the Northwest. James Wilson, compiling a volume on birds and animals "remarkable for their beauty, scarcity, or peculiarity," found Douglas constantly at his elbow with hints and suggestions. Even Professor Hooker didn't escape when Douglas journeyed up to Glasgow to visit him.

Hooker was then engaged in writing the most important work of his career, his *Flora Boreali-Americana* (The Botany of the Northern Parts of British America). Dr. Richardson had sent him specimens from his most recent expedition, as had Thomas Drummond, botanist of the Franklin party. Hooker had waited patiently for Douglas's material, which was to provide the main attraction of his book, but had not pressed for it as long as Douglas seemed busy with his own literary efforts. He wasn't surprised when he learned that work on the journal had come to a standstill (he had confided to Dr. Richardson that Douglas "has much in his head but is totally unfit for authorship"), and perhaps the invitation to visit Glasgow was at least partially inspired by the hope of persuading the failed author to release his botanical material. If so, the hope was to be disappointed. Douglas was not yet ready to give up his dreams of literary fame.

On a social and emotional level, however, the return to Glasgow was a great success. There was a warm reunion with the Hooker family; hugs from all the children and a new baby, born while Douglas was away, to be displayed and admired. The two older boys had grown considerably; the eleven-year-old Joseph, one day to be a world-famous botanist, seems to have had special rapport with Douglas. Even though it was winter, they went together to a pond so Douglas could demonstrate fishing techniques he'd learned among the Indians of the Columbia.

Then there was the triumphant return to the botanic garden; the vigorous handshake from Head Gardener Stewart Murray, the awed looks from recent employees who found it difficult to believe that this celebrated stranger had once been a workman like themselves on this very ground. There was also much that was new and exciting in the experimental plots. Through Hooker's ability to attract a flow of rare and unusual plants, and Stewart Murray's ability to grow them, the Glasgow gardens had been turned into a unique study center for the horticulturalists and botanists of Europe.

Though he did not yet wish to surrender his botanical material, Douglas insisted on making another contribution to Hooker's book. The professor was preparing a map of the territory that his *Flora* was to cover, and Douglas helpfully inserted two peaks in the Rockies that he had climbed during his trip back across the continent with the Hudson's Bay Express. He had named them Mounts Brown and Hooker, and had estimated their heights at 16,000 and 15,700 feet, respectively.

Privately, Hooker had grave doubts about these figures, suspecting that they were "egregiously overrated," but he could hardly contradict Douglas, the man who'd done the climbing. So Mounts Brown and Hooker appeared on his map and subsequently—since it's the nature of map makers to copy other people's maps—on maps throughout the world as the two highest peaks in the Rockies. It was not until well into the present century that the two summits were scientifically remeasured and their altitudes deflated to 9,156 and 10,782 feet, with Mount Hooker proving the taller.*

Such spectacular errors call for some explanation. How could Douglas have been so far off in his estimates? Probably because he started from the assumption that the Athabaska Pass, the platform from which he began to climb, was 11,000 feet above sea level. No less an authority than the great David Thompson, explorer of the Columbia River, who signed himself "astronomer and surveyer" for the North West Company, had established the pass at that height and no later traveler had questioned it. Modern reck-

* Douglas would have been surprised but delighted to learn that the Rocky Mountain peak later named Mount Douglas in his honor was higher than either, at 11,017 feet.

oning, which reduces this altitude to 5,751 feet, also reduces Douglas's errors to more respectable proportions.

Douglas, of course, without instruments, was forced to "guesstimate" the heights of mountains during his first tour on the Columbia. But even later, when supplied with all the necessary equipment, he still seemed to have trouble with altitudes. Ironically, of all his estimates, he came closest to accurate measurement with Mauna Kea, the White Mountain of Hawaii, whose summit he was to surmount but on whose slope he was to stumble.

During the summer of 1829, Douglas formed a friendship that was to be more significant for his future than he could then suppose. Through Edward Sabine he met Captain Fyodor Petrovich Lütke, a Russian explorer on his way back to Saint Petersburg after an extended voyage of geographical and scientific discovery. Botanist for the expedition was Karl Heinrich Mertens, who had accumulated an impressive collection of specimens during the cruise.

Douglas was particularly interested in Mertens's plants from Siberia, Alaska, and the Aleutian Islands since, in many cases, they indicated the northern range of his own specimens from the Columbia. For days the men pored over the two collections, cross-checking habitats, comparing notes. As the scope of their study widened, Douglas took Mertens to Chiswick and Kew, introducing him to many of his specimens growing under cultivation. Additional hours were spent in the herbaria of the Linnean Society and the British Museum.

In the course of their talks, the men developed an idea for a novel botanical expedition. The concept was that Douglas, who had already walked across Canada, should walk across Siberia, comparing plants at similar latitudes on each continent. He would also, of course, make comparative zoological and geological observations to round out the ambitious project.

Captain Lütke was enthusiastic when he heard about the scheme and undertook to use his influence at court to get consent for the expedition. He also promised to communicate with Baron Ferdinand Petrovich von Wrangel, governor of Russian possessions in America, asking him to make all necessary arrangements for Douglas to cross Siberia once the Czar's permission had been obtained.

The three men, by now fast friends, saw the undertaking as a potential *coup* not only for science but for international understanding, and drank convivial toasts to speed its realization. But Russia was far off and the Czar's permission must have seemed even more remote. When the time came for departure, Douglas must have felt, as he bid his spirited companions goodbye, that his hopes for Siberia disappeared with their sails over the horizon.

By the summer of 1829, the boundary conference had adjourned *sine die*, leaving the question still unresolved. The American negotiators had departed for home, well pleased with the results of their dilatory tactics, leaving W. R. Hay to ponder, rather desperately, ways and means to reconvene the conference before another decade passed.

He finally decided that his solution lay in science. Americans were always boasting that their young nation, born of the Enlightenment, had turned its back on Old World shibboleths to pursue the new truths of scientific facts. All hypocritical rubbish as far as Mr. Hay was concerned, and at last he saw an opportunity to hoist the Yankees by their own canards. He would give them scientific facts, more than they bargained for; he would scientifically explore the disputed Columbia area from end to end, fixing the exact location of promontories and landmarks in degrees of latitude and longitude, measuring mountains and geologically determining their composition, studying river currents, charting bays, plumbing harbors; he would accumulate these scientific facts and many more, publishing them eventually in a White Paper which he'd send to every capital of a civilized nation, including Washington, D.C. The facts should speak for themselves: British scientific enterprise and concern for the Columbia region, in contrast to American ignorance and neglect.

It was a brilliant concept. International opinion alone could force the Americans to a reconsideration of their territorial claims. W. R. Hay's main concern now was to avoid alerting them to his plans prematurely, as would surely happen if usual procedures were followed and a scientific expedition to the Columbia was formally announced by the government. The survey must be carried out efficiently but quietly by qualified Hudson's Bay Company officers and by trained personnel sent from England ostensibly to serve under them. And certainly the situation was made to order for a traveler like David Douglas, who was free to roam the country for any scientific purpose without arousing suspicion.

Before making up his mind, the undersecretary called Captain Sabine to his office and asked a question. Could Douglas be instructed in the use of observational and surveying instruments within the few weeks before the Hudson's Bay Company's annual ship departed for the Columbia? When Sabine answered affirmatively, other elements began to fall into place. The Horticultural Society suddenly found that it wanted to send Douglas back to the Pacific Coast after all, and for even more ambitious explorations; the Hudson's Bay Company again offered transportation and hospitality throughout its chain of trading posts.

Douglas wrote to Hooker in high excitement about all this. He was not only to return to the Columbia, but also "make known the vegetable treasures of the Interior of California, from the northern boundaries of Mexico, near the head of the Gulf," and even to journey across the desert to investigate "the botanical productions of the Rio Colorado." His salary was to be £120 per annum (approximately $3,000 in modern terms) and he was to be supplied with so many scientific instruments that his was clearly "not the journey of a commonplace tourist."

There was no time to be lost in learning the use of those instruments. In a cram course conducted by Edward Sabine, Douglas was put to work eighteen hours a day, at first to supplement his failed grammar school education, then to advance toward headier stuff. He was made familiar with the rudiments of geometry, trigonometry, and calculus, and provided with books for further study during the voyage. For practice with instruments the men moved out to Greenwich Observatory, and for weeks Douglas was drilled in the use of "sextants, chronometers, barometers, thermometers, hygrometers, [and] compasses of all sorts."

Finally the day came when Sabine judged his pupil ready to be entrusted with instruments of his own. He applied to the Colonial Office for a purchasing allowance and was outraged when only £80 was advanced. Spurning this paltry sum, the captain took Douglas on a shopping spree through the laboratories and workshops of the greatest instrument makers in London. By the end of it, Douglas was in possession of equipment which, in Edward Sabine's forceful if dated slang, was "prime and bang up to the mark." It was also expensive, totaling £231 8s. 6d., or close to $5,800 in a modern equivalent. Naturally, it was all purchased on credit; Sabine knew no other way to buy.

The unpaid bills were to harass the Douglas family for years after David's death. Appeals to the Colonial Office brought curt denials of responsibility, and application to Edward Sabine was equally unavailing; in fact, he became incensed at the fuss being made over an insignificant debt.

The captain had all the attributes to go far in the England of his day, including superb health which carried him into his ninety-fifth year; he died full of honors as Sir Edward Sabine, a general of the army and longtime president of the Royal Society.

Before he sailed, Douglas packed his old Columbia journal and notebooks into boxes that he delivered to the Society's headquarters on Lower Regent Street. His specimens and field notes he wrapped separately and dispatched to Professor Hooker. Then, cleared for departure, he journeyed up to Scone to say good-bye to his mother, now a widow settling down for a long twilight attended by her two unmarried daughters.

It must have been a strained visit. Mother and son had never been close, and time and distance had long since broken any lingering threads of affection. For the mother, it must also have been a puzzling experience. The thirty-year-old man in the fine London clothes must have seemed a stranger to whom she found it difficult to relate. Her older boy, John, had tried to explain to her about David; he was famous now as a result of his travels to America to bring back plants. But the explanation probably made as little sense to the former Jean Drummond as it had to the Chinooks. Why did he have to go all the way to America for plants when there were plenty in Scotland?

Also, "fame" was probably a concept she found hard to grasp in connection with her son. Fame was for lairds and the nobility; the best that people of their class could hope for was respectability, and that came from long years of work in an educated profession, such as clerking or schoolmastering. People who left home and wandered off to other countries usually returned as failures. She could have accepted David as a failure—it was the direction in which he had been headed as a child; but to adjust to him as a success was beyond her.

For Douglas, the visit must have been a matter of endurance, lasting out a sentimental duty which he couldn't decently avoid. He took what pleasure he could in conversations with his sisters and childhood friends, mostly married now with children; and of course he returned to the palace gardens where he had apprenticed.

He also purchased Billy, the "faithful little Scotch terrier" who was to be his companion for the rest of his life. He must have enjoyed training his bouncing pet. Dangers lay before them, but also warmth and affection in the cozy light of campfires. Even as they tramped through the Scottish countryside, the dog learning the manners and obedience expected of him, Douglas's mind must have jumped time and again to other circumstances and other scenes.

By summer he would be back on the west coast of America, treading on the humus of a thousand years, looking up again into the sun-filtered spires of giant trees. Close by would be a river, running white-crested, roaring as it crashed and foamed among the rocks.

These were the sights and sounds of a distant wilderness, but Douglas eagerly anticipated his return to it. For he had found that he was better suited to cope with the hazards of nature than the frictions of civilized society.

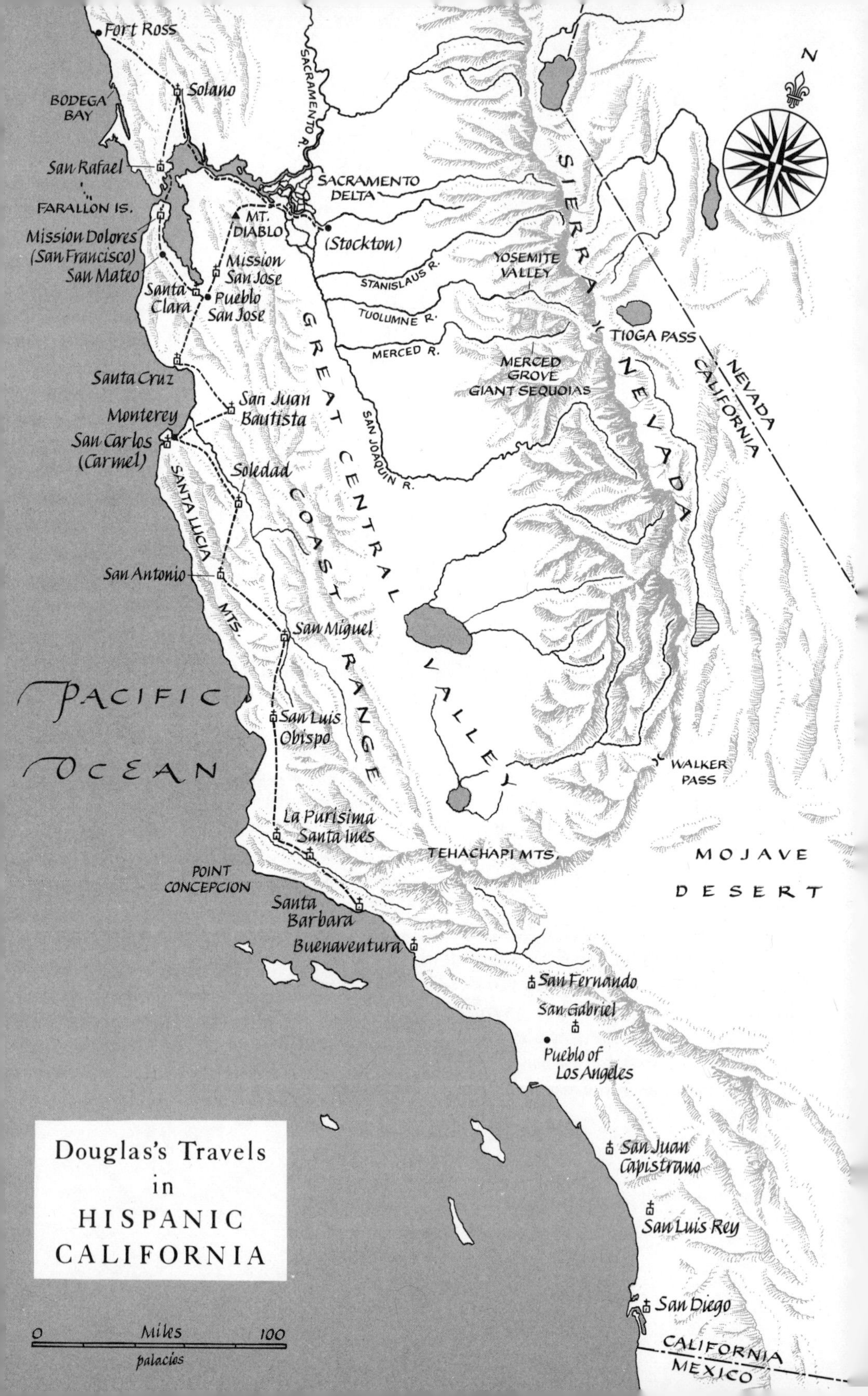
Douglas's Travels in HISPANIC CALIFORNIA
Fort Ross
Solano
BODEGA BAY
San Rafael
FARALLON IS.
Mission Dolores (San Francisco)
San Mateo
Santa Clara
Pueblo San Jose
Mission San Jose
MT. DIABLO
SACRAMENTO R.
SACRAMENTO DELTA
(Stockton)
STANISLAUS R.
TUOLUMNE R.
MERCED R.
YOSEMITE VALLEY
MERCED GROVE GIANT SEQUOIAS
TIOGA PASS
SIERRA NEVADA
NEVADA
CALIFORNIA
Santa Cruz
San Juan Bautista
Monterey
San Carlos (Carmel)
SANTA LUCIA MTS.
Soledad
GREAT CENTRAL VALLEY
COAST RANGE
SAN JOAQUIN R.
San Antonio
San Miguel
PACIFIC OCEAN
San Luis Obispo
WALKER PASS
La Purisima
Santa Ines
TEHACHAPI MTS.
MOJAVE DESERT
POINT CONCEPCION
Santa Barbara
Buenaventura
San Fernando
San Gabriel
Pueblo of Los Angeles
San Juan Capistrano
San Luis Rey
San Diego
CALIFORNIA
MEXICO
0 Miles 100
palacios
N

9

LAND OF THE GLOWING SUN

THERE IS A FAMILIAR RING TO DOUGLAS's account of his first botanizing adventure in California, late in December, 1830. "The first plant I took in my hand in full flower was *Ribes Stamin*[*e*]*um* [*Speciosum*] (Smith)," he tells us, "remarkable for the length and crimson splendour of its stamens, a plant not surpassed in beauty by the finest Fuchsia, for the discovery of which we are indebted to the good Sir Arch. Menzies in 1779." We are reminded of a description written on another first day of botanizing, on the Columbia in April, 1825: "On stepping on the shore *Gaultheria shallon* was the first plant I took in my hands."

But though the phrasing is similar, the tones of the two accounts are very different. Douglas is no longer the young enthusiast, so thrilled at touching the beautiful *Gaultheria* that he "could scarcely see anything but it." In 1830 he is resolutely academic. Once he takes the fuchsia-flowered gooseberry in his hand, he loses no time in supplying it with a Latin tag, naming the describing botanist (Sir James Edward Smith), advancing a few banal observations of his own, then climaxing the pocket lecture by playfully elevating the plant's discoverer to knighthood.

What has happened to transform the eager collector who left England the year before, anxious to prove himself worthy of his second chance, into this pontificating academician?

Possibly it was a role he slipped into more or less by accident during

RIBES SPECIOSUM (*fuchsia-flowered gooseberry*)

Originally collected by ship's surgeon-botanist Archibald Menzies, this handsome shrub was rediscovered near Monterey by David Douglas who, still unfamiliar with California seasons, was startled to find it in full bloom in midwinter. He described it as "a plant not surpassed in beauty by the finest Fuchsia."

the voyage out. As the Hudson's Bay Company's *Eagle* moved across the oceans and around the Horn, Douglas studied the ponderous books, charts, and tables with which Edward Sabine had supplied him, and seized every opportunity to familiarize himself with his equipment, sighting heavenly bodies and shooting azimuths at all hours. The ship's officers, impressed by the superb instruments, never dreaming that Douglas could be a rank beginner practicing elementary exercises, seem to have concluded that he was engaged in important though confidential scientific work, possibly connected with global measurements. It goes without saying that Douglas, given his yearning for recognition, did nothing to discourage such speculations; in fact, it's likely that he tried to enhance the respect in which he was held by silences and cryptic smiles whenever the subject of his work came up.

By the time the ship reached the Columbia, Douglas's reputation as a special scientific emissary of the government—the very reputation that Colonial Undersecretary W. R. Hay had taken such pains to avoid—was so well established that an old acquaintance, welcoming him back to Fort Vancouver, was impressed by the fact "that his stature as a disciple of science had greatly increased."

The same acquaintance can write "his noble countenance, and agreeable hearty manner [were] unchanged," but a newcomer to the fort, who had never seen Douglas before, gives a different picture, describing him as "a fair florid partially bald-headed Scotsman of medium stature gentlemanly address about 45 years of age." On the whole, Douglas would have approved of this portrait, though the exaggerated condition of his thinning hair might have piqued his vanity; otherwise the description, even to the overestimation of his age by some fourteen years, fitted in with the impression of substance and dignity that he was trying to create for his new role.

Fort Vancouver had changed since Douglas last saw it. For one thing, it had moved, relocating closer to the river. Substantial docks had been built out into the current, providing berths for oceangoing ships, a necessary adjunct to the establishment since the export of lumber and grains to Alaska, California, and the Hawaiian Islands had begun to take up the slack caused by the declining beaver trade.

The population of the fort had also increased, the roster of officers more than doubling. The new enterprises necessitated an influx of young gentlemen to command the ships and supervise work at the lumber mill and on the farms. If the Company had suffered a decline of income in the transition to new projects, it wasn't apparent at the officers' mess. Multicourse meals were served on Lowestoft china, prime Italian wines were sipped from crystal glasses, and Cuban cigars had supplanted pipes with after-dinner coffee.

In keeping with his status as a scientific notable, Douglas was seated at table close to Dr. McLoughlin; he was also assigned quarters in Dr. McLoughlin's house and, as a final tribute, was provided with a manservant.

His name was William Johnson. As with Samuel Black, A. R. McLeod, and Dr. McLoughlin himself, Johnson had been acquired in the merger of 1821. But unlike those other distinguished Northwesters, there had always been some doubt as to just where Johnson's talents lay. He had been carried in the Company's books originally as a "cow-herd," but later had been transferred to the category of "trapper." * In fact, he had trapped with McLeod's brigade through the Umpqua country when the sugar pine was found and, though Douglas makes no mention of him in his journal, he must at least have known Johnson well enough to find him unobjectionable.

It is not known if Johnson had ever served as a personal servant before, but it's certain that Douglas had never been so served. The situation may have had awkward aspects at first, but in time they worked out a relationship, though probably more as companions than as master and his man.

In late June, 1830, Douglas took off upriver, accompanied by his man, his dog, and a full pack of scientific equipment. He outdid himself in surveying along the way, fixing the positions of islands, bends in the river, rapids, and falls. When he reached Fort Walla Walla, "he was immediately busied" with more observations, since "in that portion of the Columbia, there being scarcely ever a cloud or speck in the sky, his astronomical work advanced surely and rapidly."

Then, after this show of scientific enterprise, Douglas seems to have lapsed into laziness. He spent much of the day within the adobe walls of the compound escaping the heat, venturing out only in the evenings to amuse himself by capturing lizards with horsehair snares. His host was George Barnston, a young clerk enjoying his first command, Chief Trader Samuel Black having left Walla Walla for a northern post a couple of years previously. Douglas was not tempted to travel farther up the river since John Warren Dease, the tea drinker at Kettle Falls, of whom he was genuinely fond, had died that spring. So he idled away his time, part of his lassitude deriving from his reluctance to undertake a necessary journey—into the Blue Mountains to search again for *La Grande Ronde*. The vision of sunlight and flowers had never left him, and his former unsuccessful efforts to find the valley had only strengthened his determination to succeed the next time. Yet now, with the opportunity at hand, he hung back. It was almost as if he had built his hopes so high that he was afraid to risk another failure.

* Among other employments, Johnson claimed to have been a sailor in the War of 1812. During his Hudson's Bay Company days it was presumed he had been a British tar, but later, living in Oregon under the Stars and Stripes, he recalled that he'd been a gunner on *Old Ironsides*, a memory not substantiated by United States Navy records.

But finally he bestirred himself. Barnston supplied a string of five horses—three to serve as mounts for Douglas, Johnson, and a guide, and two as pack animals to carry the formidable loads of baggage and equipment. In mid-July the cavalcade headed south.

Douglas approached the Blue Mountains down the spine this time, hoping to work from valley to valley until he found the center of the range. As the horses left the arid flatlands and climbed into the forests dotted with rain-washed meadows, he saw again with delight the plant discoveries from his former visits, the lupines and pentstemons and especially the yellow and purple peony he had named for Robert Brown.

The expedition moved steadily and uneventfully in a southeasterly direction, pausing occasionally while Douglas climbed a ridge to get his bearings. On one such excursion he must first have seen the smoke rising in the distance, though he suppressed the information when he rejoined his party. But he could only postpone the inevitable; soon enough the Indian guide would be sniffing, scenting the smoke and climbing a hill to confirm his suspicions. Then the debates would start: the guide flatly refusing to go farther, Johnson hemming and hawing but clearly reluctant to journey on into hostile Indian country.

Perhaps Douglas climbed one more mountain before they turned back. From its summit he might have glimpsed a valley set like a grail in a circle of hills; if so, it was as close as he ever came to seeing the legendary, sunlit land of the *voyageurs.*

In August, back at Fort Vancouver, Douglas had a rapturous reunion with Alexander Roderick McLeod. Much had happened to the dashing chief trader since they had last met, most of it bad. In fact, his reputation was currently under such a cloud that he was glad to accompany Douglas on an expedition into the Cascade Range in order to get away from Fort Vancouver. As he watched Douglas botanizing, or assisted him in surveying work, or as the men hunted together to provide meat for the camp pot, McLeod must have found ample time to relate his misadventures.

His troubles had started with a tragedy. In January, 1828, the Hudson's Bay Company clerk, Alexander McKenzie, Chief Concomly's son-in-law and Douglas's companion on his second trip to the Chehalis River, was murdered by Clallum Indians in Puget Sound.

Dr. McLoughlin had reacted promptly to the news; he mustered and armed a contingent of clerks and trappers and dispatched them, under McLeod's leadership, to arrest the guilty tribesmen and bring them back to Fort Vancouver for trial. Something went awry; instead of making arrests, McLeod's men attacked the Clallums, burning two villages and shooting down twenty-five men, women, and children.

Strangely, Dr. McLoughlin, usually a humane and merciful man, was not disturbed when he first heard what had happened; apparently he considered the punishment justified as a deterrent to further Indian assaults. It was only when the directors in London responded with horror to the overkill that he began to see the matter in a different light. That revelation, however, was a year off, owing to the snail's pace of communications, and meanwhile a new situation arose.

It was precipitated by the misfortunes of an American—Jedediah (Jed) S. Smith, trapper, trader, frontier adventurer. In July of 1828, Smith and a party of eighteen men, driving horses from California to sell in the Columbia territory, were set upon by the Umpqua Indians. Fifteen men were killed in what came to be known as the "Jed Smith Massacre"—a misnomer, since Smith and three others escaped to make their way to Fort Vancouver.

Naturally, Dr. McLoughlin didn't take to heart the slaughter of Americans as much as he had the death of his own man, but he sent a brigade under McLeod into the Umpqua country with the primary purpose of trapping beaver but with the secondary goal of looking around for Smith's property and retrieving it where possible. McLeod, never too enthusiastic about trapping, reversed the priorities and returned to Fort Vancouver with many of the American's horses but few beaver skins.

Dr. McLoughlin was annoyed by this failure, but his faith in McLeod's leadership was not yet completely shaken; at least, he was willing to trust him once more with a venture that might prove of the utmost importance to the Northwest fur trade.

Jed Smith, trying to repay the doctor's kindness, had told him about the fabulous beaver populations to be found along the Sacramento River in California, especially at the delta where the waters fanned out into myriad inlets before falling into San Francisco Bay. The prospects were intriguing but hazardous; a brush with the Mexican authorities, who jealously guarded California against trespassers, could lead to an international incident. The doctor brooded over the matter for a year before sending a brigade south through the Umpqau mountains and into the Sacramento Valley. When it was too late to recall the expedition, he heard London's opinion of the blundering incompetent and murderer he'd sent in command—Alexander Roderick McLeod. The doctor must have passed sleepless nights thereafter.

But he needn't have worried—at least for a while. McLeod led his brigade unerringly through the difficult mountains; he built boats and floated his flotilla without loss down an unknown river, trapping merrily along the way. He reached the delta and set up winter quarters * at about the location

* So substantial that they endured for years, giving rise to the witticism that H.B.C. stood for Here Before Christ.

of the present city of Stockton. Here he received a visit from the Commandante of the San Francisco Presidio, who urged him strongly to cut short his travels in California. With diplomatic alacrity McLeod agreed. He returned upriver, still trapping with gusto, until he regained the mountains where his horses were waiting.

Then began the series of disasters that seemed inseparable from any McLeod undertaking. Somehow the brigade was trapped in snow and ice, the horses died, the furs—a great haul of 2,400 beaver skins—had to be cached, and members of the expedition finally straggled back to Fort Vancouver, cold and starving, every man for himself.

That was the end as far as Dr. McLoughlin was concerned. Since then, McLeod had been living on borrowed time, unsure whether he would be reassigned or dismissed from the Company. The directors in London had blocked his promotion to chief factor and his career outlook seemed hopeless.*

No doubt Douglas sympathized with his old friend, but he must have been disappointed on his own account that the trip up the Sacramento River had turned out so badly. He had hoped to journey to California overland, botanizing along the way, but in the present disposition of things he realized that there would be no more brigades south for a while. Meanwhile, winter was coming on and there was no point in passing it, housebound, at Fort Vancouver.

He left the Columbia late in November on the Hudson's Bay Company schooner *Dryad*. He had been warned that the Californios lived well, so he prepared himself, as his purchases from the fort's supply room show. Packed in his valises were two pairs of Nankeen trousers, one fustian coatee and two bernagore [*sic*] silk handkerchiefs; stowed in the hold were ten pounds of Hyson tea, sixty pounds of sugar, nine gallons of Madeira wine and nine gallons of brandy. In case he might run short of knickknacks, once in the country, he took one hundred and fifty silver Spanish dollars to jingle in his pocket.

It seems sad that Douglas, who took so much from California in hospitality and reputation, thought so little of it. He lived for almost two years in that smiling land, enjoying the high tide of a civilization that has been variously and enviously called the Pastoral Paradise, Dominion of the Dons, and Eldorado of the West. He moved among men who were generous to a fault and among women whom even sophisticated observers considered

* It was to revive. After five years of banishment up the Mackenzie River, he received his promotion. He was also to receive the honor of a river named for him—not the great artery of his imagination, but a respectable tributary of the Sacramento, unfortunately misspelled, with his usual bad luck, as the McCloud River.

superlatively beautiful. Though a foreigner, he was allowed to roam freely over most of the vast province and was assisted in every way to collect the seeds that were to revolutionize the look of European gardens. Yet, in spite of all, he could still write after sailing away, "The Mexican territorial government as applied to California is abominable . . . the secular part of the community is so-so, some good and many bad," and could even complain—a crowning irony—that the country was better fitted for geological investigation "than for botany."

It must be said that Douglas was in a distracted frame of mind when he expressed these views on California, but they reflect substantially what he had come to feel about the country. It was too different; the climate was too uniform, the living too easy. The gaiety of the people, at first so attractive, came in time to seem hedonistic, rubbing against the grain of Douglas's provincial ethics. In the course of nineteen months, he traveled full circle, ending with a violent distaste for the fruits of Lotus Land.

It had been different when the *Dryad* first dropped anchor in Monterey harbor, "a short distance from the sandy white beach" beyond which, in the words of another visitor, "rose pine-covered peaks to prick a sky as brilliantly blue as the bay." The small town, the capital of California, which Douglas saw that December 22, 1830, consisted of about forty one-story adobe houses, most with red-tiled roofs, some white-washed, "dotted about, here and there, irregularly . . . a pretty effect." The population, including the military at the Presidio and the families on surrounding ranchos, numbered about seven hundred. According to Richard Henry Dana, who had ample opportunity to make comparisons when he arrived before the mast some four years later, "Monterey . . . is decidedly . . . the most civilized looking place in California."

The arrival of a ship in the harbor was a rare enough event to bring out crowds, and, as Douglas was rowed ashore, he was curiously observed by many of Monterey's inhabitants. Among them was a man who stood out because of his height and general bearing. He was William Edward Petty Hartnell, a transplanted Englishman who was to be Douglas's host for much of the next two years.

Hartnell, born in Lancashire and by now thirty-two years old, had been a resident of Monterey for the past eight years. He had come originally as a purchasing agent, to buy cattle hides and tallow for a British firm, but, as Americans entered the field and competition grew stiffer, he had begun to look around for additional sources of income. His great ambition was to be appointed British consul and chargé d'affaires for California, and, to prove his value, he had for years turned his house into a business and hospitality center for visiting Englishmen. His services were appreciated in Whitehall though, typically, nothing had been done to recompense him by way of an appointment, while guests, of whom Douglas was merely the latest, continued to be sent him.

Hartnell owned one of the larger houses in town. It was a single-story adobe like the others, but was built on a U plan, which allowed for easy expansion by adding room cells at the open ends. Hartnell's own family was still small, since he had been married only five years, but there were usually in-laws and guests to be housed. Douglas was pleased when he found two rooms assigned to him, one to serve as a separate study and workroom.

Douglas soon learned that Hartnell was considered the spokesman for the foreign community in Monterey, numbering some fifty in all, mostly of American or British origins. His leadership derived partly from his connections (he had married into one of the prominent California land-grant families), partly from his fluency in languages (he spoke Russian, French, and German as easily as Spanish and his native tongue), but mostly because of his honesty and very real ability. He was widely respected, and even the one debilitating flaw in his character was generally forgiven if not condoned.

Hartnell was, of course, a convert to Catholicism and a Mexican citizen; otherwise he couldn't have married and held property or even enjoyed legal rights to life and liberty. We may surmise that renouncing king, country and the Church of England may have caused him pangs, but it was the only realistic thing to do if he wanted to make his way in California. Nor was William E. P. Hartnell—or Don Guillermo Arnel, as he became known after his conversion—alone in coming to this conclusion. Apostasizing—"leaving one's conscience at Cape Horn," as the popular phrase put it—became the first law of survival among immigrants.

There was Thomas W. Doak, a native of Boston, who became Don Felipe Santiago. John Cooper, also from Massachusetts, was transformed into Don Juan Bautista Cooper, though more familiarly known as Don Juan El Manco because of a deformed left hand. The Buckle brothers, Samuel and William, English sailors and laborers, were metamorphosed for the sake of wives and property into José Manuel Boc and José Guillermo Buchel respectively. And finally, Henry Jubilee Bee, also English, was baptized Enrique Ascension, the transformation of his last name being particularly noteworthy.

Even Douglas, though he had no thought of renouncing King William IV and the Presbyterian Church, was caught up in the tide of Hispanic nomenclature and, before he had lived long in Monterey, became known as Don David El Botanico.

Douglas soon recognized that seasonal growth patterns in California were different from those in other lands. Grasses and annual flowers, especially, tended to burst into leaf and bloom in response to rains, no matter when those rains occurred during the year. He had written about this in some surprise to Dr. Hooker: "Early as was my arrival on this Coast [December] Spring had commenced." He was anxious to start his botanical journeys, particularly to the south, before the flush of green faded from the hills, but he found that the authorities were in no hurry to grant him the necessary travel permit.

He could understand the reason for the initial delay; Manuel Victoria, the newly appointed governor of California, didn't arrive in Monterey to assume office until late in January. But when, subsequently, excuses and evasions in the matter of the permit continued, Douglas fumed and fretted and began to build the image of an "abominable" government.

In actual fact, Governor Victoria, looking at it from his point of view, was perfectly justified in delaying Douglas's departure from the capital. He had taken office at a critical time. The former Spanish, now Mexican, province of California, long famed as "the most peaceful and quiet country in the world," was beginning to show the strains of a conflict that in a few years was to tear it apart. Already the polarization of factions had begun. Victoria had learned, for instance, that his predecessor as governor, José Maria Echeandía, instead of returning to Mexico was loitering in the pueblo of Los Angeles. Echeandía was notoriously absentminded * and if Victoria had been charitably inclined he might have assumed that the former governor had simply forgotten to go home. But he was neither that charitable nor naïve. Los Angeles was a rallying point for the Paisanos, the landless native sons who were plotting to reapportion California and who numbered as allies British and American adventurers. For all Victoria knew, the Scottish newcomer to Monterey, who was pressing so hard for a travel permit, was in a hurry to join fellow conspirators in the south. Better to let him cool his heels for a while until more was known about him.

Douglas, finally resigned to delay though continuing to huff at official red tape, made good use of the time imposed upon him. He had the whole Monterey Peninsula to explore, a territory so rich in its flora that for twenty years after Douglas had skimmed off the cream it continued to make botanical reputations. He was particularly pleased with the novel conifers, always a Douglas specialty. Staring him in the face every time he stepped outside the Hartnells' door were clumps of a dark-green Christmas-tree-shaped pine, which was to prove so fast growing in cultivation that it became an important lumber tree in many parts of the world. We know it, appropriately, as the Monterey pine *(Pinus radiata)*. Later, in the hills to the south, Douglas was to find the digger pine, which he named for an old friend *Pinus Sabini-*

* So much so that, when faced with signing a state document, he is reputed to have asked, "What's my name?"

ana, and subsequently, the bigcone pine, which he named for a new friend he was to make—*Pinus Coulteri*. Of the seventeen known species of pine on the Pacific Coast, Douglas discovered seven, an amazing record.

Also to be found on the peninsula were coast redwoods *(Sequoia sempervirens)*, though he was to save his comments on that giant until he had seen the impressive groves farther north. He also identified for the first time the Monterey cypress *(Cupressus macrocarpa)*, the twisted, storm-tossed tree whose wind-sculptured shape has come to stand as a symbol for the area.

Douglas makes no mention of the Monterey cypress in any botanical note or letter that has survived; if he collected seeds, they were lost in transit or in the upheavals that were shortly to convulse the Horticultural Society of London. In fact, another collector, Karl Theodore Hartweg, is usually credited with introducing the tree in 1848. How then do we know that Douglas even spotted it eighteen years earlier? Because he sent back to Edward Sabine the latitude and longitude of a promontory which he called "Cipres Pt," thus establishing his claim to the conifer and naming the rocky bluff on which it still grows most picturesquely.

Douglas also visited the mission of San Carlos Borromeo in the Carmel Valley, a few miles across the hills from Monterey. San Carlos (or Carmel Mission, as it was and is universally called) was the second establishment founded (1770) in California by the saintly Franciscan, Father Junípero Serra, who was charged with saving Indian souls for God, while his escorting soldiers took possession of Indian lands for their Spanish king. Carmel was a long trip overland from San Diego, the first established mission (1769), but it was readily supplied by sea, an important consideration in founding the early Franciscan outposts.

Majordomo at Carmel was Father Ramón Abella, well into his sixties and one of that select but dwindling group of padres who had arrived in California prior to 1800. Even in 1798, however, he was too late to know either Father Serra or his second-in-command, Juan Crespi, though he had known and prayed with Fermín Francisco de Lasuen, the last of the great founding trio, all now interred in the sanctuary of San Carlos near the altar. Father Lasuen alone lived long enough to see the glory they had all dreamed of rise above their resting place—a stone church, seventh in a line of buildings that started with a crude cabin of logs and progressed through various structures of adobe and tile.

Douglas may have had an opportunity to exercise his newly brushed-up Spanish on Padre Abella, but it is probable that he first visited the mission in the company of Hartnell, in which case he would have left conversation to that accomplished linguist. There was much to see in the great cloistered quadrangle that abutted on the church. Workshops, stretching in ordered rows, housed Indian converts trained in leather-working, weaving, carpentry, and even iron-forging. They industriously turned out shoes, cloth, furniture, and utensils for field and household. Fires burned continuously in the kitchens, heating the stews and gruels that constituted the main mission fare. In the center of the courtyard, Douglas saw the great vats and presses for the manufacture of olive oil and wine, and the huge black caldrons in which lard was rendered into tallow for candle-dipping.

He was, of course, familiar with self-sufficient communities from his residence at Fort Vancouver, but there were important differences. The Hudson's Bay craftsmen, recruited in Scotland or Canada, were masters at their trades; the Indians at Carmel, as at the twenty other missions in California, had had to learn almost overnight new skills and attitudes which suddenly plunged them into an industrial-mercantile age centuries removed from their native cultures. In addition, the Hudson's Bay Company forts were provided annually with goods, supplies, and even luxuries from England. The missions received no such support. From 1821, when revolutionary forces overturned Spanish sovereignty, to 1825, when the new Republic of Mexico undertook to resume authority over its provinces, California had been a forgotten land. The missionaries learned that they must not only sustain themselves and their Indian charges, but feed and clothe the entire Spanish-speaking population as well, including, of course, the idle and parasitic soldiers.

Only the willingness to work and the adaptability of the converts, called "neophytes," saved the day. For years they accepted mission discipline, trying hard to be good children so that they might be rewarded in heaven. They even learned to play musical instruments, homemade violins and oboes, and to sing, harmoniously if uncomprehendingly, two-part hymns in Latin.

But of all the arts to which the neophytes were introduced, they took with greatest enthusiasm to horsemanship. Many of them developed phenomenal skill at roping, cutting, herding, and other accomplishments of the range. Indian vaqueros made possible the great cattle industry, both at the missions and at private ranchos, which enriched and enlivened the last decade of the Golden Age. Their feats with the reata or lasso in steer-wrestling and riding wild horses were legendary, and their contests laid the base for modern-day rodeos.

As he rode back to Monterey with Hartnell, Douglas may have been curious as to why, in this empty country, the mission was so far removed from town. It was because of the Presidio, Hartnell would have told him. Soldier populations and neophytes could not coexist. The noncoms raped the Indian girls and the officers impressed Indian men into servitude. At San

Diego as well as here, the missions had been forced to move away from the Spanish settlements.

Douglas could have learned something else from Hartnell that would have surprised him. The mission padres felt no responsibility for the spiritual welfare of the town communities. Their duties started and ended with the saving of Indian souls. If townspeople wished to come to the mission for confessions, weddings, christenings, and the like, they were welcome, but the padres could not be expected to travel out to towns and ranches, even to administer last rites.

Also, the Franciscans, superbly educated themselves, turned a deaf ear to pleas that they open up schools for settlers' children. If schools and the comforts of religion were wanted, let the people send for their own teachers and priests.

But somehow teachers and parish priests never came; or, if they did, they escaped back to Mexico as soon as possible. The Royal Chapel, built for the garrison and civilians of Monterey, for want of religious occupancy was used as a storeroom, and generations of Californios grew up so unlettered that even the most successful, who acquired enormous cattle herds and ranges, were unable to sign their names to legal documents.

For years it seemed that the civilian population, possessed by an indolence called "California fever," was also resigned to ignorance, but it turned out to be not so. The resentment against the padres was building, and, when it finally broke, it took a violent form.

"The season for Botanizing is not more than three months," Douglas wrote Professor Hooker. "Such is the rapidity of Spring that plants (like on the table-lands of Mexico and the platform of the Andes in Chile) bloom only for a day. The intense heat sets in about June when every bit of herbage is dried to a cinder."

As February advanced and still no passport was issued, Douglas grew increasingly desperate. It was turning out to be a dry spring—he recorded less than an inch of rain—and he had visions of encountering wilted woods and withered meadows if his journeys through the countryside were delayed much longer.

Hartnell tried to expedite matters but found the new governor a stubborn marinet and not to be pushed. He succeeded, however, in obtaining permission for Douglas to accompany him as far as San Juan Bautista on his regular business circuit of the missions. Douglas was authorized to visit Santa Cruz as well; it was within a day's travel of the capital and also in the loyal north, removed from the conspiracies of Los Angeles.

In mid-February the men set out, first riding down the spine of Monterey Peninsula, then crossing the vast salt marshes, *las salinas*, amid clouds of wheeling ducks indignant at being disturbed. They reached high ground by midafternoon and finished the thirty-mile journey before dusk.

Many missions on view today have been so "restored" that it's hard to tell where restoration leaves off and the real article begins, but San Juan Bautista is an exception. Refurbishings and replacements, necessary because of years of earthquakes and neglect, have been so skillfully contrived that not only the physical appearance but the "whole original mission aspect" are still much as Douglas saw and experienced them in 1831. Even the paintings on the ceiling of the church (some executed by that ship's carpenter Thomas W. Doak, who disappeared into Don Felipe Santiago, and others by nameless neophytes) have been reproduced from sketches and memories still extant at the beginning of this century. All that seems missing is the acrid scent of fennel, which used to be scattered on the tile floor to discourage fleas from hopping on worshippers.

Douglas looked forward to meeting Padre Felipe Arroyo de la Cuesta, the valetudinarian in charge of the mission, about whom he had read a spicy account written by British officers passing through San Juan Bautista a few years previously. Father Arroyo had royally wined and dined his visitors and bedded them down comfortably for the night, but the next day mysterious things began to happen. First, the horses seemed to have become too lame for travel; when the Englishmen offered to buy new mounts, their saddles were inexplicably missing. Through one ruse or another, the officers were forced to remain at the mission an extra day. The reason became apparent when the padre, who had been continuously plying them with wine, undertook to convert them, "reading . . . innumerable lectures in refutation of the Lutheran and Calvinistic doctrines, and in favour of the Pope's supremacy." It took all night before he realized he was wasting his time, and then he allowed his guests to travel on.

It is unlikely that Father Arroyo made any attempt to convert Douglas; by this time he was probably resigned to the fact that heretics embraced the true faith only when they had something material to gain by it. For his part, Douglas may have wondered what had become of the padre's liberality with wine, since he saw nothing but goat's milk and water on the night of his arrival. But he understood better, the next day when, after Hartnell's departure for other missions, wine bottles suddenly appeared at table. The flaw in Don Guillermo's character was alcoholism, which made him subject to morbid fits and depressions. Whenever possible, the padres conspired to protect him from himself.

Left alone to make his way in halting Spanish, Douglas found that his Franciscan host possessed a lively and inventive mind. For instance he learned that the padre had developed an early version of the alarm clock;

run by water, it could ring a bell at his bedside. Among speculations that interested him was the origin of the word "California," another place name of uncertain derivation. The padre offered a choice of theories: the name came either from a Latin description of the climate, *Calida Fornax* (hot furnace), or from the prevalence of resinous pines in the country (*colofon* being Spanish for resin, California became the Land of Resins).

Apparently Father Arroyo had never heard of Queen Califia, a potentate who ruled an island rich in pearls, gold, and precious stones "on the right hand of the Indies." This mythical monarch and her fictional island—spelled, to the letter, "California"—appeared in a romance published in Spain around 1502. The theory is that, fifty years later, stout Cortez applied the name to Lower California (then thought to be an island) in the hope it would bring him luck in finding another treasure of Montezuma. Unfortunately for the theory, neither Cortez nor his co-conquistadors could read, and it is doubtful that the priests with his party would waste their time on romances.

There's a tempting speculation that the name derived from the Greek words *kalli ornis,* meaning "beautiful bird." Only the central letter "f" is missing to make the thesis viable; but, sadly, etymologists tell us that a central "f" is *never* developed in the course of language usage.

Which leaves us with one last straw to clutch at for the most likely derivation but, fortunately, it seems substantial. The historian Hubert Howe Bancroft, taking last testaments from pioneers a century ago, heard of an Indian phrase from Baja California which he judged might register on Spanish ears as *kali forno*. It means "high ground" or "native land."

Douglas rode out from San Juan Bautista, heading north and west through the Coast Range for Mission Santa Cruz, accompanied by an armed and mounted dragoon. This slouching soldier, or his counterpart supplied by the squad room of every mission he visited, was to become a familiar companion of Douglas's travels, and always bitterly resented.

In this Douglas was irrational. He had never objected to the *voyageurs* and Indians Dr. McLoughlin had insisted he take along on the Columbia; in fact, he had welcomed the company and protection they represented. But not so in California. Even while he acknowledged his guard's usefulness in guiding him across sometimes trackless country, he insisted on regarding him primarily as a hostile presence, foisted on him to spy. In one way or another, Douglas was going to make good his charge of an "abominable" government.

Mission Santa Cruz, now vanished in any but symbolic form, was then a cluttered compound of adobe buildings surmounted by a bell-towered church. In charge was Father José Joaquín Jimeno, young, energetic, born in Mexico, different from the padres Douglas had thus far met. "A more upright and highly honorable class of men I never knew," he was to write of the Franciscans. "They are well educated; I had no difficulty from the beginning with them, for, saving one or two exceptions, they all talk Latin fluently, and though there be a great difference in the pronunciation between one from Auld Reekie * and Madrid, yet it gave us but little trouble."

No doubt Padre Jimeno was one of the exceptions to the fluent Latin-speaking men among whom Douglas was pretending to count himself. The founding missionaries and their immediate successors, all born in Spain (whose "cradles stood in Spain," as the famous phrase ran), were either dead or on their last legs, and with them was to pass the tradition of Latin as the universal tongue of scholars. The new men, born and trained in Mexico, were taught only enough Latin to conduct services and were sneered at by the old padres as "curates" rather than priests.

Also, the new Franciscans, reared in republican times, tended to hold liberal views, and questioned such matters as their right to flog Indians, issues that had never for a moment troubled their predecessors. And there was a further difference between the generations. The older men prided themselves on their fair, European complexions. Though we have no description of Padre Jimeno, his brother, also Mexican born, was "dark in person," a characterization which alone would have prejudiced Douglas against him.

Whatever he may have thought about his missionary host, there is no doubt about Douglas's reaction to the great forests of redwoods that crowded the hills behind Santa Cruz. They took him back to his first days on the Columbia, when he had gazed with awe into the towering canopies of the Douglas firs and scrambled with a tape to measure fallen giants. "The great beauty of California," he wrote Professor Hooker, "is a species of *Taxodium*, which gives to the mountains a peculiar—I was going to say an *awful*—appearance, something which tells us we are not in Europe . . . I have measured them frequently two hundred and seventy feet long, thirty-two feet round, three feet from the ground!! Some few I saw upward of three hundred feet long, but none of a greater thickness."

He was aware that Archibald Menzies had seen and reported on the tree, as indeed had two botanists before Menzies, but no one had yet introduced the plant to cultivation and Douglas was eager for that honor. "I have fine [botanical] specimens," he told Hooker, "and seeds also!" What became of these seeds, we don't know; but the collector who was finally credited

* Edinburgh. It is interesting to note that even then certain cities had reputations as air polluters.

with introducing the coast redwood (in 1846) was the same Karl T. Hartweg who rediscovered the Monterey cypress.

With his bandoliered companion in tow, Douglas returned to the capital early in March to find that a celebrated guest had taken up residence at the Hartnells' during his absence. She was Doña Concepción Argüello, the heroine of pastoral California's one authentic romance and a legend in her own time.

Though past forty when Douglas met her, Doña Concepción was still renowned for "her expressive and pleasing features" and still "distinguished for her vivacity and cheerfulness." Though the gray habit of the Tertiary Order of Franciscans, which she wore these days, may have inhibited to some degree her "shapeliness of figure" and "a thousand other charms" which had so entranced admirers when she was sixteen, it is doubtful that, in spite of her best efforts, she had been able to dim the "love-inspiring and brilliant eyes and exceedingly beautiful teeth" which had devastated Nikolai Petrovich Rezanov, ambassador and plenipotentiary of the Czar, twenty-five years earlier in San Francisco.

Hers was a Cinderella story, wonderfully romantic, though, unlike the original, ending in tragedy. It was a tale to rend the heart and blind the eyes with tears and, in bookless, illiterate California, it was told and retold till voices gave out.

Concepción—or Concha, as she became known to legend—was a simple comandante's daughter in the Presidio of San Francisco when she was seen and promptly loved by Count Rezanov when his ship put into harbor. The feeling was mutual, but more than the usual number of obstacles arose to obstruct matrimony. Because of his rank, he must obtain an emperor's permission to wed; because of her religion, Rome must be appeased since her lover was not of the Faith. Nothing daunted, the count sailed from San Francisco vowing to resolve all difficulties and return for his bride within a year. Unhappily, within months, he was thrown from a horse and killed.

Concha grieved and remained faithful, a constancy to be severely tested as innumerable suitors, foreign as well as domestic, pressed her for marriage. At one point she cut off her hair to cool the ardor of her admirers; when that didn't work, she considered cutting off her lustrous eyelashes—until a priest forbade it. She took vows and adopted the habit of a religious, but even that didn't immediately solve the problem. A delegation of caballeros called on her father demanding that he "restore his daughter to life." It was only when the doughty old warrior chased them from his doorstep that the stricken swains of California finally gave up.

In recent years, Doña Concepción had been living in Santa Barbara, dedicating herself to good works with such devotion that she had won the title of "La Beata." It was typical of her kindness that she had come up to Monterey that spring to assist Mrs. Hartnell through what promised to be a difficult accouchement.

It is likely that Douglas felt more at ease with the friendly Franciscan sister than with any woman he had ever known. Her gray habit and matter-of-fact manner seem to have dispelled the uneasiness he usually felt in the presence of the opposite sex. Perhaps, too, her humor injected a surprising element into their relationship. Concha had always possessed a sense of the ridiculous; to an American suitor, who pointed out that she could save his soul through Catholic conversion if she married him, she is supposed to have replied, "But, Señor, with the great comforts of my religion you won't need any small comforts I might provide." So it is possible that, from the first, she adopted a teasing attitude toward the serious-minded Scot which, when he recovered from his surprise, he found delightful. And certainly no one but Concha could have invented the nickname by which he became known in the Hartnell household: Saint Francis, because of his love for flowers.

At the end of April the long-awaited passport was issued and, though no limitations to travel were officially spelled out in it, events clearly indicate that Douglas was forbidden to journey farther south than Santa Barbara, the last bastion of loyalty upon which Governor Victoria could depend. Within a week, at the head of a small cavalcade of pack animals and spare mounts, Douglas was on his way down the Monterey Peninsula. He spent the first night at Mission Soledad, characterized by a jaundiced contemporary as "the gloomiest, bleakest, and most abject-looking spot in all California," then hurried on across the Santa Lucia Mountains and reached, by the following evening, the Mission San Antonio de Padua in what was called the Valley of the Bears.

Presiding at San Antonio was Father Pedro Cabot, noted among the padres for his Old World urbanity, "a fine, noble-looking man, whose manners and whole deportment would lead one to suppose he had been bred to the courts of Europe, rather than the cloister." He belonged to that category of mature, authoritative yet kindly men, such as Professor Hooker and Dr. McLoughlin, with whom Douglas seemed always to strike up immediate rapport, so it is no surprise to find him spending almost a week at the mission, when his original intention had been to hurry south in advance of the warming sun, which was already beginning to wither spring vegetation.

As Father Pedro conducted his guest past the mission orchards and rode with him across the oak-dotted pastures that supported six thousand cattle, ten thousand sheep, and nearly a thousand horses, he might have been forgiven the sin of pride. For almost three decades he had been vicar of this vast estate, in years of plenty now but once at a time of disaster. The 1812 earthquake, which damaged nearly every mission of the chain, had struck San Antonio a paralyzing blow, reducing it virtually to rubble.

Father Pedro was not wanting in the will to start reconstruction, but the hands to help him were few. Some of his converts had been killed when the buildings collapsed; many more had run to the hills in panic. The padre prayed for guidance and received what seemed a cryptic answer: his salvation lay in music and feasting. The father understood; he set his remaining neophytes to singing and cooking. Giant stew pots were kept continually simmering, the savory aroma drifting back into the mountains. Slowly, one by one at first, then in family groups, the fugitive converts returned, unable to resist the music and the scent of food.

Father Pedro conducted a great mass of thanksgiving, and the work of restoration went on apace. Earlier in 1831, the year of Douglas's visit, a critical observer had found the mission "in the most perfect order; the Indians cleanly and well dressed; the apartments tidy; the workshops, granaries and storehouses comfortable and in good keeping."

Yet despite this gratifying picture, Douglas found Father Pedro greatly concerned about the health of his charges. His neophyte population had diminished, in ten years, from a thousand souls to less than seven hundred. The problem was not unique to San Antonio; death rates exceeded birth rates throughout the mission system at about 10 percent per year.

It was a puzzling phenomenon. Past epidemics of smallpox and scarlet fever had taken an expectedly heavy toll, but, recently, minor infections—chicken pox, whooping cough, measles—considered little more than nuisances in Europe, were carrying off both adults and children. And there were other indications of a population under stress. Few children were born to mission converts and, of these few, the majority were males. In epidemics, too, more girls and women died than boys and men.

Father Pedro was trying to correct this imbalance between the sexes by an innovation. He was encouraging unmarried girls and widows to confide in him their choice of available males; he hoped that, by arranging marriages between such partners, he could increase the birth rate through the connubial enthusiasm of the wives. But if this scheme failed, he didn't know what to try next. Some missions had hinterlands where unconverted Indians still abounded, but San Antonio had no such potential recruiting ground. The remaining aborigines of the Santa Lucia Mountains were too few and too stubborn to be worth the effort of salvation.

Once more in the saddle at the head of his small cavalcade, Douglas progressed on down the Coast Range, spending a night at Mission San Miguel Arcangel and, the following day, riding through expansive country where grassy meadows patched with wildflowers rolled out from beneath majestic oaks. As afternoon wore on, herds of cattle replaced browsing deer, and finally Mission San Luis Obispo came into view, a columned church flanked by an adobe compound nestled in a lush valley, still another to be called "Valley of the Bears."

The bears referred to were grizzlies, so called because their gray-tipped hairs gave a frosted or grizzled appearance. Bears were numerous in old California, as is evidenced by the state flag (a grizzly prowling on a field of white) and memorials peppered throughout the countryside. Hardly a county is without its Bear Creek, Bear Hollow, Bear Mountain or Bear Gulch. There were, of course, other animals in the territory too, and deer, elk, beavers, and cougars have been duly celebrated in place names, but never to the extent of bears, and for a very good reason: A chance meeting with a bobcat could slip from the mind, but an encounter with a grizzly, weighing half a ton, nine feet tall when reared on its hind legs, fangs and claws bared and screaming like a banshee, left an indelible impression.

Father Luis Gil y Taboada, the padre at San Luis Obispo, had once been noted for his medical skills, though his reputation was now under a cloud. In fact, he had been stopped by his superiors from further practice, an injunction which the padre considered both unjust and opposed to holy purpose. He never tired talking about it and it's unlikely that Douglas escaped hearing his complaints.

Father Gil, interested from boyhood in the healing arts, had been delighted to find in California a veritable drugstore growing at his front door. In addition to the traditional European remedies, he had inquired into and adopted Indian medicinal herbs, such specifics as yerba santa for catarrhal conditions, cascara sagrada as a cathartic, and a wonderweed called *Canchalagua* as a cure-all for almost everything, including snakebite.

Of course, if drugs failed to cure a patient, Father Gil resorted to phlebotomy. There was nothing unusual in this since it was widely believed that bloodletting was the only sure remedy for an unresponsive medical problem, but in time it was noted that the padre took to the knife more frequently and with greater ambition. To make matters worse, it was rumored that his favorite patients were women, especially pregnant women, and the expectant mothers among his neophytes lived in terror of him. When called before a church tribunal to answer charges that he had performed unnecessary Caesarean operations, Father Gil boasted of the souls he had snatched from the devil and saved for God by getting at the infant while there was still life in it, and was outraged when he was forbidden all further medical exercises.

ERIODICTYON CALIFORNICUM

Yerba santa (holy plant) was highly regarded by Hispanic Californians as a remedy for colds, grippe, asthma, and other respiratory ailments. The crushed, aromatic leaves and stems were used as an inhalant and also brewed into a tea for internal use.

SATUREJA DOUGLASII

Yerba buena (good herb) was much prized as a remedy for indigestion and, taken in the form of a minty tea, as a febrifuge for reducing fevers.

CENTAURIUM VENUSTUM (*canchalagua*)

This member of the gentian family was credited with a wide range of curative powers–the wonder drug of early California. The bitter tonic concocted from its leaves was used as an antiseptic, a specific for catarrh, and a remedy for snakebite.

RHAMNUS PURSHIANA

The bark, called sacred bark (cascara sagrada), was stripped from this large shrub or small tree and processed into a laxative, an essential physic in a land where the diet was overwhelmingly meat.

It was recognized that he was sick, of course, and should be retired from active service. But the shortage of priests made that impossible. The founding Franciscans had planned on two fathers for each mission to share duties and relieve loneliness, but it had never worked out that way. The revolution in Mexico crippled the Franciscan order, and few priests could be spared for remote California. At the time of Douglas's visit, the twenty-one missions were manned by twenty-six padres, too slim a margin adequately to cover the emergencies of sickness and death. As long as Father Gil could stand up and perform the sacraments, he was needed.

Douglas continued south, passing successive nights at Missions Purisima and Santa Ines. Then his horses labored over the San Marcos ridge and he saw again the wide expanse of the Pacific, lost to him since Monterey. The small cavalcade moved cheerfully through the remaining miles to Santa Barbara and the home of Don José de la Guerra y Noriega, comandante of the Presidio and father-in-law of William E. P. Hartnell, who was to be Douglas's host during his stop in town.

Rarely has a man so dominated a community as Don José did Santa Barbara, and it was a status he had achieved through wisdom, courage, kindness, and other admirable qualities. In a country where law courts were nonexistent, Don José was the Solomon to whom disputes were brought for settlement; in a land without banks, his word was enough to insure credit; in an age when there was no organized charity, his purse was always open for the needy, and he had long since proved his talents as a soldier. When the pirate Bouchard threatened the town, Don José marched his handful of soldiers in a classic maneuver around and around a hill, making their ranks seem to stretch endlessly.

Physically this remarkable man was unprepossessing. He is described as of short stature, stout, flat-nosed, and ugly, but in spite of his looks, or lack of them, he succeeded in fathering four girls famous throughout California for their beauty. It was said that he had repeatedly been offered the governorship of the province, but he refused to renounce his allegiance to the Spanish king and the land of his birth in order to take office under the Mexican republic. It was no wonder that with all this accumulated prestige, ordinary mortals passing La Casa Grande, Don José's home in the center of town, lifted their hats respectfully.

Santa Barbara was known as a gay place. Monterey was considered the dull habitat of sobersided business men and politicians, but Santa Barbarans were always looking forward to fiestas and celebrations, and Douglas's arrival must have been a prime excuse for staging a fandango. In a matter of hours the courtyard of La Casa Grande would be converted to a dance floor, musicians summoned, and messengers dispatched to the ranchos

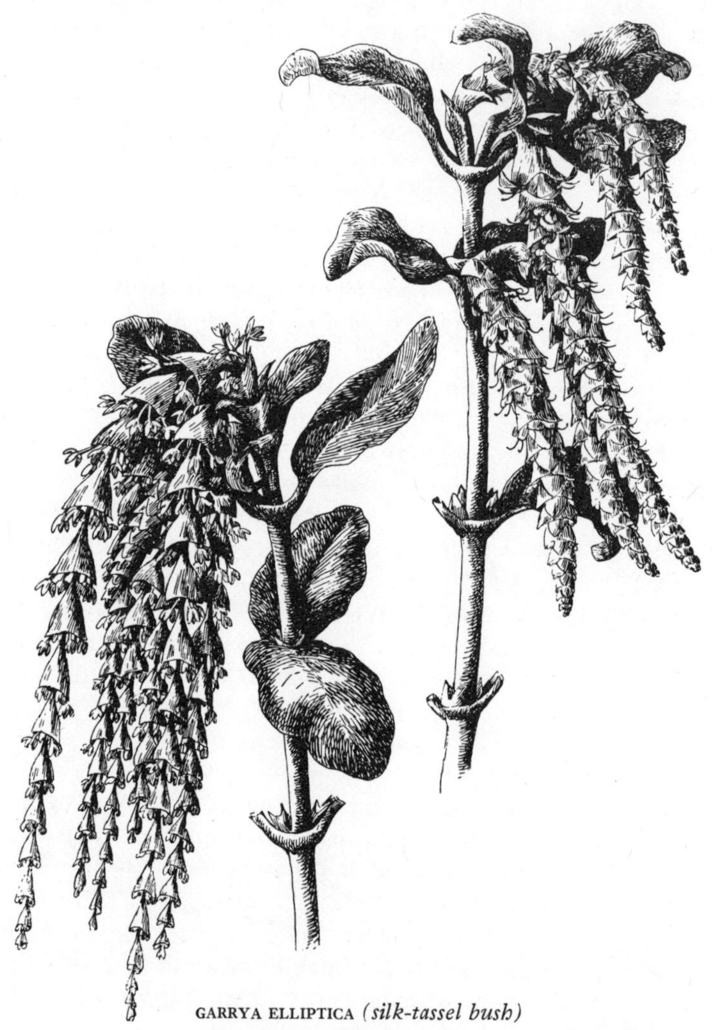

GARRYA ELLIPTICA *(silk-tassel bush)*

This shrub or small tree is a botanical oddity in that it comprises an entire plant family of only one genus. It was discovered by Douglas in California in 1832 and named for Nicholas Garry, deputy governor of the Hudson's Bay Company, in appreciation for "kind assistance."

to call home the young caballeros for the party. Music began in the afternoon, Indians and mestizos occupying the floor; but by evening, when the lamps were hung and the candles lit, the *gente de razón* * would take over.

Douglas, no dancer, would have dancing forced upon him by the laughing señoritas, as had many leaden-footed Britishers before him. Actually, demands on the male weren't taxing. He was only expected to jog in place, more or less in the rhythm of the guitars, with his hands clutched behind him and his eyes fixed on his partner's feet. That's where the excitement lay, small feet popping with incredible lightness and dexterity in and out under voluminous skirts. If a girl was "called out"–given the floor to herself–she would sweep and swirl as the music played faster and faster, propelling herself by unseen feet like a floating bird. Or perhaps castanets would appear on another dancer's fingers, urging on the music with crackling whips of sound. Bystanders, everyone in town, would crowd the edge of the floor, clapping to the music, uttering cries of ecstasy when carried away by a dancer's performance.

A fandango developed a life of its own, and a good one could last for four or five days and nights or longer. The music continued endlessly; so did the dancing, though individual musicians and dancers might rest for a few hours, or conceivably even eat. There was little drinking. Wine, inseparable from mission hospitality, was rarely served in private homes except at weddings, birthdays, and similar ceremonies, and then was consumed in moderation. On isolated occasions Californios might become high, but habitual drunkenness was considered an affliction of foreigners.

Douglas, in spite of himself, or perhaps with enthusiastic cooperation from himself, seems to have been caught up in the Santa Barbaran spirit. "The ladies are handsome," he wrote George Barnston, his friend on the Columbia, "of a dark olive brunette, with good teeth, and the dark fine eyes, which bespeaks the descendents of Castille, Catalan [Catalonia] or León." He finished the letter with unexpected archness: "They (sweet creatures) have a greater recommendation than personal attractions. They are very amiable. On this head I must say, *Finis*, otherwise you will be apt to think, if ever I had a kind feeling for man's better half, I left it in (Calida Fornax) California."

Much significance has been attached to this passage, and efforts have been made to link Douglas with one or another teen-aged beauty. Nonsense. It was because he wasn't attracted to any debutante in particular that he was able to describe the entire spectrum of Santa Barbara girls in such

* Literally "people of education." But since there were few such in California, the term soon took on class and racial overtones.

glowing terms. For Douglas, there was not only safety in numbers, there was positive salvation.

Douglas always had a soldier's compulsion to carry out orders to the letter. In his instructions from the Horticultural Society, he had been authorized to travel south to the Mexican border, which was still some two hundred miles south and east from Santa Barbara.

He may have consulted Don José as to the advisability of breaking his verbal commitment to Governor Victoria and proceeding on down the mission chain to San Diego. If so, Don José undoubtedly advised against it. Douglas would find himself *persona non grata* when he returned to Monterey; he might be forced to leave the country by the next ship, and thus never get a chance to explore the rich territory to the north. Don José was no botanist, but he knew that plants stayed greener in the summer up there. Soon now, in southern California, the heat would settle down in full force, shriveling and parching everything.

The climatic argument would have been a telling factor with Douglas. He had already noted, as he passed Point Concepción on his southward journey, that he had moved into the latitudes of dry desert winds. Given his fair complexion and his sensitivity to heat, he must have suffered on the days he botanized through the dry arroyos above Santa Barbara. The prospect of even hotter weather would have dismayed him and possibly hastened his return northward.

As he rode back over the mission trail, he may have consoled himself with the thought that he had collected everything of importance in the entire southern California area, and as far as shrubs and herbaceous plants are concerned, this was generally true. But there were still two noted species of pine he might have encountered below the Tehachapi ridge. Possibly he wouldn't have seen the one-leaved piñon *(Pinus monophylla)*; it is a denizen of interior uplands and is generally located well east of the road Douglas would have traveled. But the other one, on a hill above Del Mar, within a few miles of Mission San Diego, would have jumped out of the landscape at him because of its unusual shape and coloring.

It is a pity he didn't find this tree, if for no other reason than the privilege of naming it; because the collector who did find it named it in honor of Dr. John Torrey, the man who, years before in New York, had called David Douglas "such a liar."

10

EMPIRE OF HIDES AND TALLOW

DOUGLAS RETURNED TO MONTEREY IN late June, 1831, and almost immediately started out again on his travels. His objective was to complete a botanizing tour of northern California before the heat became too oppressive but, owing to delays and distractions of one kind or another, he was still out in the sun in August.

And he suffered. In spite of "Bernagore" silk handkerchiefs swathed around his face and neck, water bags evaporating across his saddle and noontime siestas in the shade, Douglas found the oven conditions of high temperatures and low humidities almost unendurable. "In no part of the world have I experienced such a dryness in the atmosphere," he complained, "nor can I call to my memory having read of greater. Even the deserts of Arabia and Egypt, the plains of Sin and Ispahan in Persia, I mean the driest places on the globe, when satisfactory observations have been made, are more humid than California. Often when the thermometer Fahr. stands at 80° or 100°, 30° or 40° of dryness is by no means unfrequent." And writing to Professor Hooker in a similar vein, "The heat is intense . . . 129° not infrequent, which, if I mistake not, is not exceeded in Arabia or Persia . . . How I lament the want of such majestic rivers as the Columbia."

These complaints directed against California's famous, pre-smog weather make us wonder if they could be based on realistic fact, if the intensity and dryness of the sun's rays could have been greater in Douglas's time than in our own. For climates do change significantly in something less than millenniums. We know, for instance, that some six hundred years ago, summers turned so dry in Europe that the Rhine could be forded at Cologne. Shakespeare's injunction to "fear no more the heat of the sun" makes little sense in today's chilly England, but four hundred years ago the country was experiencing a continental cycle of hot summers and icy winters. Even in America in the last century, bitter cold invaded New England for an extended period known as the Little Ice Age, culminating in the "Year of No Summer" when three feet of snow lay on fields until well into June.

So, factually, California *could* have been experiencing a period of unusual weather during Douglas's visit. We know that it wasn't, however, because of evidence compiled by other contemporary observers. Captain F. W. Beechey of H.M.S. *Blossom*, for instance, kept accurate and detailed logs on temperatures and humidities at the California ports he visited between 1825 and 1828, and his statistics show the climate to be just about what the National Weather Service finds it today.

But what about Douglas's specific observations, his statement that temperatures of 129° (Fahrenheit) were "not infrequent," or his readings of "30° or 40° of dryness" (presumably meaning percentage of relative humidity in our terms)? Professional weathermen can only assume "instrument error," possibly incorrect techniques of thermometer exposure that lead to the temperature of overheated instruments being read rather than degrees of free-flowing air.

Perhaps the hottest weather Douglas encountered that summer was at Mission San Jose, and the highest temperature ever officially recorded there is 108° F., a far cry from 129° F. As for his regretting the absence of "such majestic rivers as the Columbia," he had San Francisco Bay and the entire Pacific Ocean along much of his travel route, supplying relative humidities of from 65 to 85 percent, almost exactly paralleling the readings at Fort Vancouver.

It is hard to avoid the conclusion that Douglas's sufferings from the California climate were largely psychological and occurred because his instruments told him it was both hotter and drier than it was. In contrast, Captain George Vancouver, more expert at reading thermometers and hygrometers, was able to enjoy in the western Eldorado "a climate remarkable for its salubrity."

To his surprise, Douglas met that summer a dynamic and enterprising leader, a heroic figure on a level with Chief Trader Alexander Roderick McLeod and Captains John Franklin and Edward Sabine. The surprise lay in the fact that Narciso Durán wore the tonsure and habit of a Franciscan priest, though there were those who said that the padre of Mission San Jose de Guadalupe was a priest by accident only, the accident of a humble birth in Spain, leaving the church as the only avenue down which an ambitious boy might travel.

Father Durán was in his middle fifties when Douglas knew him, of medium stature, somewhat stout, blue eyed and fair complexioned. It was rumored that he had asked for service at San Jose because of its reputation as the most troublesome of the missions. Situated inland, east of San Francisco Bay, it was within easy striking distance of the gentiles (or "wild" Indians) who inhabited the tules * of the Sacramento delta. Isolated cattle-stealing incidents had occurred since the mission's founding in 1797, but under the leadership of Etanislao, a renegade Christian convert, Indian raids became organized and damaging.

Padre Durán personally led the expeditions sent to capture Etanislao and, though the will-o'-the-wisp rustler evaded the lumbering Spanish troops for years, he was eventually betrayed and killed, leaving his name to posterity in a county and a river, both called Stanislaus.

When these hostilities—pastoral California's closest approach to Indian warfare—were terminated, Padre Durán directed his energies toward building up the prosperity of his mission. He doubled and tripled his heads of livestock, cultivated orchards, and especially rejoiced in his vineyards, whose grapes were distilled into the famous "clear" brandy of San Jose.

Douglas arrived at the mission at a time of great excitement, for the *matanza*, the annual slaughter of cattle for hides and tallow, was about to commence. Some twelve thousand head ranged the rolling hills, though by day few were to be seen. After four months without rain, the color green had virtually disappeared from the countryside; grass was to be found only in sheltered arroyos and draws, where the cattle congregated beneath stunted oaks. In addition to harvesting hides to fulfill contracts with the trading ships, the *matanza* served a useful purpose in thinning out the herds to numbers which could be successfully pastured through the rest of the dry summer.

Early on each dew-drenched morning, Douglas rode out with his host to view events. The padre had a favorite hill, a command post where, under a wide-brimmed hat, his habit tucked around his saddle exposing bare legs, he could watch, like a supervising general, the drama of the day unfold. The vaqueros would already be at work, nudging cattle out from the shelter of ridges and trees, herding them into the open. The cattle moved accommodatingly and without suspicion at first. Since birth they had become accustomed

* Literally, bulrushes or swamp grasses, but the meaning was extended to apply to any wilderness area that might provide cover and concealment (pronounced toó-liz).

to roundups: first for branding, then for castration of the males (leaving a ratio of twenty-five cows and heifers for each bull), and finally the weekly rousting and jostling as fifty bullocks were cut out from the herd to be driven to the mission to butcher for the pot.

Slowly, systematically, about a thousand cattle would be driven into the first of the "holding" areas. Then, in groups of twos and threes, prime bullocks were separated and herded over a hill toward the killing ground.

As the oncoming steers smelled blood and saw the sights with which the turf was littered, they drew back, started to bawl, and often panicked and ran. But they did not get far. A vaquero would reach down from a speeding horse, grab the animal by the tail and, with a jerk, upend it. Reatas would loop, roping and holding, until the butchers could run in for the kill. Animals that stood frozen with fright, making no attempt to run, would simply be lassoed and thrown for their throats to be cut.

A busy scene ensued when the last animal was down. Long knives ran red, flaying carcasses and staking the hides out to dry. Pockets of fat would also be removed from the corpse to be taken back to the mission for rendering into tallow. In all, a little more than a hundred pounds of saleable substance (worth about four dollars) would be removed from a thousand-pound animal.

At the end of a day's work, the killing ground looked like a ghoulish battlefield, littered with red and skinless objects. Already the vultures would be circling and, by sundown, the coyotes, bobcats, and cougars would be howling and snarling in the hills, waiting for dusk before they crept down to gorge themselves.

Later would appear the monarchs of the California wilds—grizzly bears, one of the few animals to share with man a relish for both meat and vegetables. On the final night of the *matanza*, the vaqueros would lie in wait to lasso at least one large and active grizzly for the most famous of the country's spectator sports, a bull-and-bear contest. The animals, matched in size and strength, though opposites in fighting methods, were tied to each end of a lariat and released in a corral. For a while each tried to escape on its own, only to be jerked back by the other captive. This led to irritation, anger, and, finally, attack.

If Douglas ever saw a bull-and-bear fight, it was that summer at Mission San Jose. If he didn't, he was in good company. Most visitors to Hispanic California heard of the epic struggles, but few claim actually to have seen one. In some quarters the bull, because of its greater strength and sharpened horns, was favored. Others swore by the bear, whose wily tactics included grabbing first the nose and then the tongue of the bull, and hanging on till strangulation resulted. In any case, blood flowed, spectators screamed in wild excitement, and the consensus as to which animal won most often is completely unimportant.

Douglas claimed he got on well with the padres because he avoided religious and other controversial topics. As a result, "I had no bickerings about superstition, no attempts at conversion or the like, the usual complaints of travellers; indeed so much to the contrary, that on no occasion was an uncharitable word directed to me."

Presumably, at San Jose, as elsewhere, Douglas tried to avoid discussing California politics as being controversial, but it is likely that as he and Father Durán sat over the famous clear brandy he heard a great deal more than he cared to on the subject. For the padre, serving his first term as president of the missions, was deeply troubled, as were all Franciscans, by threatening winds blowing in from Mexico and up from southern California.

The threat lay in a legal measure called "secularization." It was an enlightened concept that had been basic to the founding of the missions, the theory being that the missionaries held the lands in trust for the Indians. Some day, when the native populations were suitably Christianized and educated, the lands would be returned to them. When that day came, the missions would shrink down to parish status as the padres gladly handed over the vast church holdings to civilized and responsible Christian converts.

The padres had no quarrel with the principle of returning the lands; the only question was *when*. They pointed to the dependency of their neophyte charges, and suggested that in another fifty years—certainly not before—secularization might be considered.

The Paisanos * and their political allies in Mexico disagreed. The time for secularization was now. Admittedly, few neophytes were yet politically or socially mature in a European sense, but that difficulty could be surmounted by appointing *administradors* to look after their interests. The *administradors*, to be chosen from among the Paisanos, would fulfill the same functions the padres had in the past, managing the landholdings and negotiating the sale of hides and tallow. The Franciscans could stay on with the reorganized communities if they wished, but strictly in a spiritual capacity. They would have no hand in running the estates.

We can well believe that Douglas shared Father Durán's indignation at these proposals, which seemed ill-disguised attempts to confiscate church property. No doubt the Paisanos stood condemned in his eyes as monsters of greed and hypocrisy. Yet had he known the full background, he might have reserved at least some understanding for the native sons who, in the course of

* Literally, native sons or children born in the country. In California politics, however, the name came to signify a certain group of aggressive, land-hungry freebooters.

years, had developed certain legitimate and deeply felt grievances against the padres.

One, known as "bounding," was essentially ecclesiastical land-grabbing. The boundaries of mission properties were extended till they touched each other, thus squeezing out the possibility of private lands between. Mission San Jose was a case in point. It was off the beaten track and therefore had no value as a link in the mission chain. It had no harbor or other natural feature to justify its existence strategically. What the site had very obviously to recommend it was thousands of acres of rich pasture, and for that reason the church had laid claim to the property. What is more, the boundaries in the course of years were expanded almost to touch the lands of Mission Santa Clara to the south and west, thus seriously hampering the growth of the pueblo of San Jose, lying between.

But in addition to practical grievances, the Paisanos had more subtle reasons for their resentment of the padres. They never forgave the Franciscans for their attitudes of superiority, for their pride in their fair Spanish complexions, for their contempt of mestizos, and their indifference to the schooling and spiritual welfare of the colonists and their children. These resentments rarely found open expression, but were deep-seated nonetheless. The padres had proved themselves "bad" fathers and the sons were bent on punishing them.

There is no other way to account for the ruthless harassment that ensued. Within two years of Douglas's departure from California, the political pressure the Paisanos were able to bring to bear on the Mexican government prevailed, and secularization became the law of the land. The native sons, now in the saddle, set about administrating (often a euphemism for looting) the former mission properties with a free hand, reducing the neophytes to serfs and the padres to paupers.

But impoverishment wasn't enough; the Paisanos undertook a campaign of humiliation, trying to drive the Franciscans out of the country. Some who refused to leave literally died of starvation at their altars. And finally, in a climactic gesture, just before the onrushing Yankees swept away the degenerated civilization of the dons, the Paisanos auctioned off crumbling mission churches above helpless priestly heads.

San Jose, alone among the California missions, maintained a satisfactory level of neophyte population. This was because Father Durán took direct action whenever disease or runawayism reduced his flock; he sent soldiers and converts into the hinterland to capture wild Indians as replacements. For years he had delighted in leading such expeditions himself, but, since being censured for "excess of zeal" in conversions, he now judged it best to let others carry the Cross of Jesus into the tules.

There is reason to believe that such an evangelizing party left Mission San Jose that summer and that Douglas went with it, in which case he would have traveled north through the present Livermore Valley, then on past Mount Diablo to the Sacramento delta. We must infer this route since we have no written record. Douglas's California journal was afterward lost in a near-fatal accident on the Fraser River, and no letter survives detailing his journey to Mission Solano, his next known stop. But we can be reasonably sure of his itinerary because of some ingenious detective work performed by botanists in recent years.

Among the seeds and specimens that Douglas sent back to the Horticultural Society that summer, three plants would seem to pin down the road he traveled. The first, a member of the mustard family, is found only in the inner Coast Range, which would indicate a route well east of San Francisco Bay. The second plant, a mariposa lily, is found only in two locations, one of which is in the vicinity of Mount Diablo. But the third specimen, another mustard, seems to settle matters beyond dispute since it is found *only* on the slopes of Mount Diablo.

Given these clues, another becomes immediately evident. Douglas had a compulsion to climb mountains, so Diablo, visible from a great distance, would have attracted him. At 3,849 feet, it wasn't much of a peak by his standards, but it was there, the highest summit in the range, and undoubtedly he climbed it.

He probably took leave of the mission party at the edge of the delta, possibly in the neighborhood of the present city of Stockton. As he surveyed the tangled thickets of willows, alders and poplars intersected by sluggish canals, one thought would be dominant: he had reached the site of Chief Trader Alexander Roderick McLeod's farthest penetration south when he had brought his beaver brigade barreling down the Sacramento River two years before. Perhaps it was then that he conceived the idea of traveling farther north than he had intended in order to link up with his former journey into the Umpqua country. Douglas always yearned for continuity, for travels without gaps, for goals achieved without hiatuses or interruptions, and the prospect of climbing back into the sugar pine mountains, approaching the great trees from the south this time, must have excited his imagination.

San Francisco Solano was an afterthought among the missions, a loose link dangling at the northern end of the chain. Comparatively speaking it was brand new, founded on July 4, 1823, a date on which Douglas was crossing the Atlantic for his fruit-tree-buying mission in New York and points adjacent; also, Solano was the only Franciscan establishment which had been

CALOCHORTUS LUTEUS (*yellow butterfly lily*)
Common in northern California, both in the Sierra Nevada and the Coast Range. Douglas may have collected seeds and bulbs on Mount Diablo, where the yellow lilies especially flourish.

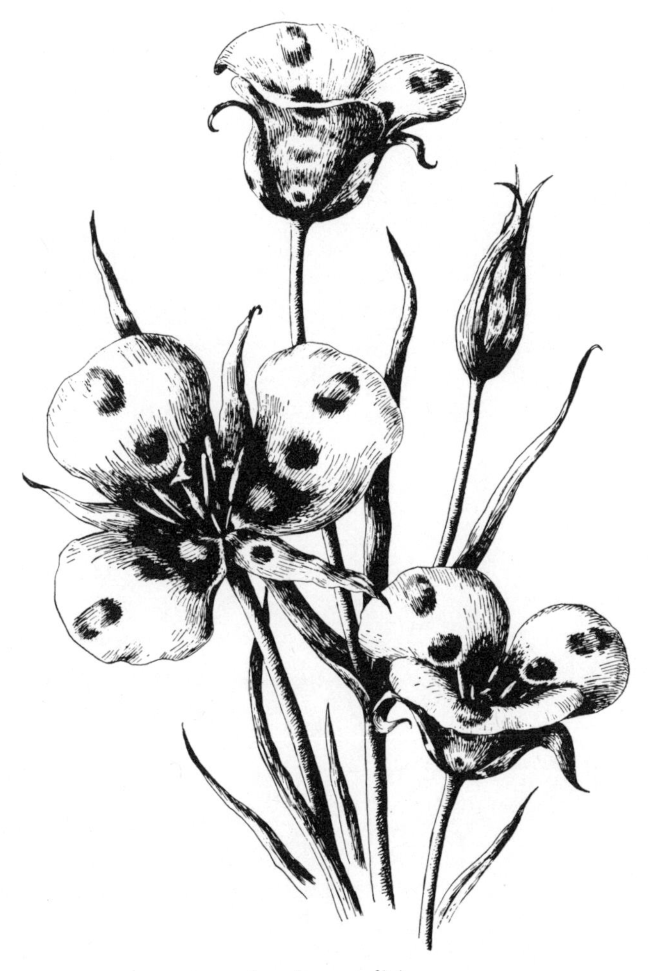

CALOCHORTUS VENUSTUS (*mariposa tulip*)

This delicate white to pale lilac blossom, patterned with reddish polka dots, was found by Douglas on a dry California hillside nodding its cupped head charmingly ("venustus" means charming) in a passing breeze.

erected not to save souls but to serve as a vantage point from which to spy upon a neighbor.

For the Russian-American Fur Company, controlled from Moscow and digging in with an increasing air of permanency on the Mendocino coast, had long worried California's governors under both Spanish and Mexican rule. Repeated demands that the trespassers quit the country had gone unheeded, and emissaries sent from Monterey to Fort Ross, the main Russian redoubt, had been politely received but evasively answered. Finally, the Hispanic authorities, too weak to move militarily, had had to take what comfort they could in the construction of a missionary observation post from which surveillance on the foreigners could be maintained.

Padre Buenaventura Fortuni, the kindly scholar who presided over Mission Solano, was startled by the questions Douglas put to him soon after his arrival. Guides and hunters for an expedition into the northern mountains? Where indeed could such people be found? Not among his Indians; he had none to spare and, besides, they were far too timid to venture out of their own valleys.

But he had a suggestion: Douglas might find the men he was seeking at the Russian garrison on the coast. For, despite prohibitions from Monterey, Father Fortuni knew a good deal about his Slavic neighbors; he had been carrying on a brisk trade with them for years, exchanging beef and grain for manufactured goods; in fact, even his mission bells had been imported from Moscow. The Russians were the best hope for the kind of enterprise Douglas had in mind.

Moving up streams and across ridges of the Coast Range, Douglas experienced what must have been for him the most enjoyable traveling conditions since he arrived in California. He was in redwood country, and he rode for miles under a canopy of the towering trees, across carpets of needles that muffled all sounds. In between the Sequoia stands were scattered groves of Douglas firs whose spreading branches and graceful droop would have carried him back to the Columbia. It was a land of rainfall and soft mists, which, no doubt, stilled his usual complaints of dryness and excessive heat.

Breaking out to the coast, Douglas reached scenery that may have reminded him of parts of Scotland—rugged headlands and windswept, almost treeless heaths. His first sight of Fort Ross must have been reassuring. It stood foursquare on a bluff above the ocean and looked, except for a minareted chapel that peeped above the stockade, much like a Hudson's Bay trading post. But this pleasing impression dissipated rapidly once Douglas found himself inside the compound.

"Undisciplined" seems to be the word most often used to describe the atmosphere of Fort Ross. The garrison consisted of about a hundred men—fifty Russians and fifty Aleutian Islanders. With the exception of a few officers and noncoms, the Russians were said to be "the sweepings of slums and jails."

It was the business of the Aleuts to hunt for sea otter along the coast, and of the Russians to grow food and raise stock, not only for themselves but to supply other trading posts in Alaska; but there seemed to be very little work of any kind performed. The Aleuts claimed there were no otters left, and the Russians had almost given up trying to farm the cold, gopher-ridden ground. So, defying the noncoms' knouts, the men drank, quarreled, and trafficked in the favors of local Indian women.

Douglas spent only one night at Fort Ross. It is doubtful that he even broached his reason for coming to Peter Kostromitinov, the commanding officer, since he had no intention of employing in any capacity the ruffians he saw about him. Giving up all thought of the Umpqua Mountains, he headed back for Solano, then traveled on south through Mission San Rafael to San Francisco Asis, better known then, and now, as Mission Dolores.

Douglas was considerably depressed by his failure to mount an expedition to the sugar-pine country. Since it was an ambitious undertaking and he seems only to have thought of it at the last moment, we wonder at his extreme dejection until we read a letter he wrote about the matter to Professor Hooker. "My principal object was to reach the place from whence I returned in 1826 [somewhere to the west of the present town of Roseburg, Oregon] which I regret to state could not be accomplished. My last observation was 38° 45′ N. [Fort Ross] which leaves a blank of sixty-five miles. Small as this may appear to you, it was too much for me!!"

We are startled by this miscalculation. The blank he speaks of is more like three hundred than sixty-five miles. The most reasonable explanation is that Douglas, and McLeod too, lacking instruments in 1826, thought they had penetrated much farther south on the sugar-pine journey than they actually had.

It is easy to smile at such gross geographical errors today when accurate road maps are available free at every gas station. But being first in a new terrain is always bewildering. There are the points of the compass and nothing else. Trees, rocks, streams, begin exactly to resemble one another, and there is no telling how far you've come because there is no telling where you're going.

Looking back from our perspective, we can consider Douglas lucky that he was spared a journey bound to be mortifying and one that could easily have ended in disaster. After all, less than three years' time had elapsed since the Jed Smith massacre, and Douglas was preparing to travel through the same Umpqua territory by the same route.

Though he didn't know it yet, he was soon to attempt a much more ambitious expedition, for which he would need all his confidence and courage unimpaired to see him through.

Douglas lingered for several days at Mission Dolores, relishing the mists that blew in through the Golden Gate, reluctant to face the high temperatures and hot winds awaiting him in the valleys to the south. But finally, swathed against the sun, with Billy trotting ahead and the inevitable dragoon bringing up the rear with a string of fresh mounts, he started down the San Francisco peninsula.

As he climbed the San Bruno ridge he could glimpse far out at sea the Farallon Islands, the most successful of all the Russian provisioning stations, where some fifty thousand tough and fish-flavored sea gulls were annually killed, split, and dried. By noon he had reached the mission stock-farm called Buri-buri, where a *matanza* was in progress.

We have the estimate of Juan Bandini, one of the Paisano leaders, that eighty thousand cattle were slaughtered that summer for their hides and tallow. It was a total that had jumped in geometric progression over the past few years because of an unprecedented demand for shoe leather in Europe and the United States. Three factors were at work: for the first time in history, shoes were comfortable to wear because a distinction had been made (as recently as 1818) between the fit of the right and the left feet; shoe manufacture, traditionally the work of individual cobblers on a custom basis, was being standardized in factories that turned out ready-to-wear shoes at lower prices; industrial populations, which had once made do with sandals, carpet slippers, and wooden clogs, were taking to leather shoes as symbols of urban sophistication.

This ever-expanding market brought California its first viable economy—raw materials in exchange for manufactured goods—and the missionaries and a few Spanish land grantees, like Don José de la Guerra y Noriega with his three hundred thousand acres, benefited hugely. Many smaller rancheros also grew rich enough to buy silver tea services, ornamented shotguns, silks, satins, and a host of other luxuries brought by the trading ships.

Tallow production, of course, swelled with increased hide shipments, but it too found its way to a market hardly threatened by glut; for throughout the world candles—made from tallow or, more expensively, from beeswax—were still virtually the *only* means of illumination. Gas, introduced in the second decade of the century, was still under suspicion, and kerosene was more than thirty years away. So from the humble cottage, with its single sputtering wick, to palaces graced by chandeliers, candelabra, and wall sconces, the terrors of the night were kept at bay solely by candle power.

Everywhere that summer, on mission lands and private ranchos, the killing went on as hides and fats were harvested, processed, and prepared for shipment. Waste materials were left to fester where they fell. The meat, rotting by the ton, overwhelmed the efforts of natural scavengers to consume it. The stench of putrefaction filled the air for miles around, and the pious, fearful of disease, crossed themselves when the odor of death entered their homes.

The road beyond Buri-buri looped back toward the bay, reaching it in the neighborhood of what is now Burlingame, at the edge of a great salt marsh. Douglas, riding close to the water, must have observed "wild geese . . . covering whole acres of ground, or rising in myriads with a clang that may be heard at a very considerable distance." At San Mateo, another Mission Dolores property, nets were spread on high poles in order to entangle geese as they flew inland to feed at dusk.

South of San Mateo, rolling park land began and continued all the way to Mission Santa Clara. Oak trees, often knotted with mistletoe, provided shade as the horses passed beneath; coveys of partridge so tame that they could be killed with sticks must have reminded Douglas of birds on the Galápagos Islands; and finally, at the end of the day's ride, he saw the never-failing mirage of Santa Clara, a spectacular illusion whereby the mission seemed to rise out of a huge lake.

Somewhere during those last miles Douglas may have observed a patch of wild oats *(Avena fatua)*. He would recognize the plant as an alien, a common weed grass of Europe, its seed probably imported by accident in straw wrapping. As yet (1831) the patch of oats would cover a few acres at most. Four years later it had crept north almost to Mission Dolores, but the stand was still confined to the peninsula. We next hear of *Avena fatua* in 1855. By then the oats had run wild, completely covering the Coast Range and the great Central Valley of California, making, in the words of one observer, "a carpet as complete as buffalo grass on the prairie."

We cannot condemn this invasive alien * out of hand. For one thing, its yellow straw gave a name to California; "Golden West" was neither applicable nor used in the Hispanic period. For another, wild oats make nutritious hay and perform well as soil binders. But we're curious about the original native grasses which preceded *Avena fatua* and were pushed out by its irresistible vigor.

Douglas didn't think to describe them, nor did a host of botanists who followed him to California. Their names, styles, and habits might have been irretrievably lost to us but for the brilliant work of recent investigators. Adobe bricks, taken from the first buildings erected by the padres in Cali-

* Wild oats, though the most visible and perhaps most famous of California's naturalized immigrants, is only one of an estimated 526 alien plants that have learned to call the Sunset State home.

fornia, were reduced to dust and examined for seeds. The results seem to indicate that the pastures on which the mission herds waxed fat were comprised of annual and perennial clovers and various kinds of bunch grasses.

We must particularly regret the disappearance of the clovers, which performed a vital function in fixing nitrogen in the soil. It was noted that, when wild oats first ran rampant, there were high nitrogren areas where stalks grew so tall that a man on horseback could be hidden by them.

Padre José Viader of Mission Santa Clara had been a giant in his prime. The story was told that when he was attacked by three marauding Indians, he knocked the heads of two together immediately and caught the third while he was trying to escape. The heathen trio were so impressed by the muscular holy man that they offered themselves up for conversion.

The padre was past sixty when Douglas knew him, and though no longer the athlete of yore, he was still impressive in a large-boned, gaunt way. He had the reputation of being reserved with strangers, adopting a stern, abrupt manner that could, however, dissolve in the glow of his famous Burgundy. For Father Viader had made his mission as noted for its wines as his near neighbor Narciso Durán had made San Jose renowned for its brandies. When Douglas wrote about California, "The wine is excellent; indeed, that word is too small for it; it is very excellent," he possibly made the amendment in consideration of the well-stocked cellars of Santa Clara.

As the men discussed mutual acquaintances Douglas must have learned of Father Viader's special regard for Doña Concepción Argüello, an interest which went back many years. The padre had known her in 1806 when, sixteen years of age and one of those "beauties . . . one may find, though but seldom, only in Italy, Portugal, and Spain," she had met and charmed the middle-aged and worldly wise widower Count Nikolai Petrovich Rezanov, plenipotentiary and diplomat of the Czar.

The padre had been called in by the Argüello family and the governor of California to help with the problems arising from the May-October romance—perplexing problems with ecclesiastical and political implications beyond the capacity of provincial officials and priests to solve. It had seemed at the time that Count Rezanov had adopted the only wise course when he left his sweetheart behind and sailed away to appeal the case to *his* Emperor, *her* King, and the Pope in Rome. But after a quarter of a century of ruminating on the matter, the padre had come to believe that the Russian nobleman's departure was a little too slick and his romantic intentions open to question.

Of course he was dead now, thrown from his horse the following year while crossing Siberia on a journey that may, or may not, have been primarily to obtain the Czar's consent to his marriage; so the question of faithfulness was academic, except for the effect of Rezanov's promises on Doña Concepción. If the padre was right, she had wasted more than twenty years grieving for a man who had cared nothing for her.

How does one prove such charges? It would be difficult in law where motives beyond doubt must be tied to actions. But the case that Padre Viader had marshaled against Rezanov was morally convincing, strong in its implications of hypocrisy and deceit.

The count had been in a tense frame of mind when his ship first arrived in San Francisco. His objective was to obtain food—and with dispatch—for starving Russian garrisons in Alaska. Within hours of dropping anchor, he had hurried to visit nearby missions where he saw cattle, wheat, dried peas, and beans—everything he needed except the necessary permission to buy them and load them into his ship. For the Spanish laws were strict; there could be no trading with foreigners. Foreign vessels were not allowed in California ports except in distress or emergency, and neither condition applied to the count's trim and seaworthy ship.

In due course the San Francisco comandante hinted that it was time for Rezanov to be moving along. He became positively insistent when a message was received from the governor in Monterey asking why the foreigner was still in harbor.

It was then, and only then, that Rezanov discovered the charms of the comandante's daughter and began laying siege to her heart. Almost immediately the atmosphere changed. Doña Concepción's parents, the mission padres, even the governor when he heard about it and hurried to San Francisco to see for himself, were enchanted by the romance. Not only could it elevate a simple daughter of California to the highest levels of European society, but it might inaugurate a new alliance between the Spanish and Russian empires. Needless to say, though Rezanov repeated it tirelessly, such a diplomatic coup would lead to honors for the Californios who had promoted it; the governor, comandante, and Franciscan priests could expect dazzling rewards. All that was necessary now was permission for Rezanov to conclude his business in Alaska so he could start immediately for Saint Petersburg.

The governor obligingly returned to the capital and Comandante Argüello looked the other way while Rezanov's ship was provisioned and sailed off, sped by the prayers of all true Californians. But Father Viader, after thinking about the matter for years, had come to see facts previously obscured by the aura of romance. The reality was that Count Rezanov, in a little more than a month, had worked his way around inflexible Spanish regulations and got exactly what he came for, a shipload of food. He had done it by the simple strategem of paying court to a local beauty and thus, by arousing grandiose notions in provincial minds, had turned her kith and kin, as well as legal and spiritual advisers, into an alliance of lawbreakers, conniving on his behalf against their own king and country.

It was not for nothing, the padre finally concluded, that Count Rezanov had been considered one of the Czar's most astute and wily diplomats.

Returning to Monterey in late August, Douglas must have observed Doña Concepción with new interest. Had she ever suspected that to her Russian lover she might have been nothing more than a pawn in a devious bunco game? Or did she believe, with the rest of California, that if Rezanov had lived he would surely have come back to carry her off to his palace in Saint Petersburg?

It was impossible to tell from her consistently bright and cheerful manner. She certainly didn't seem to view herself in the role others had invented for her as the heroine of a tragedy. In fact, there had always been those who, marveling at her lightness of spirit, wondered why, years ago, after a suitable period of mourning,* she had refused to avail herself of one of many attractive offers of marriage. Could she have anything against California men, or men in general?

Perhaps Douglas best answered that question, though unwittingly, when he observed of California girls, "They (sweet creatures) have a greater recommendation than personal attractions. They are very amiable." And amiable they certainly proved themselves to be on becoming brides. With few exceptions they lost little time in tuning up the machinery of reproduction. At a period when Indian populations were dying off in missions and native villages, the statistics of fecundity among the first families of California are formidable. Secundino Robles and María García had twenty-nine children; Tomás Sánchez and María Sepúlveda (she was thirteen at marriage) had twenty-one; J. A. Castro and Merced Ortega, twenty-six; Ignacio Vallejo and María Antonia Lugo, thirteen; Joaquín Carrillo and María Ignacia López, twelve. The list continues, with loaves no sooner out of the oven than a fresh batch was warming up.

Not to be forgotten are Douglas's hosts in Monterey, William E. P. Hartnell and María Teresa de la Guerra y Noriega. Officially they are credited with nineteen children, but late in life the old lady indignantly claimed that the count should have been twenty-five. The figures aren't

* Both Bret Harte and Gertrude Atherton, who produced considerable kitsch about Doña Concepción, suggest that she waited faithfully for decades, not knowing if Rezanov was dead or alive. This is fantasy. She would have known of his death, within a year or two, from diplomatic sources, or, at the very latest, within six years, when the Russians arrived to stay on the California coast.

important, but the pride in procreation is revealing. In fact, the spirit of the thing seems to have infected Hartnell to such an extent that he fathered at least one additional child out of wedlock.

It seems probable that Doña Concepción had no biological yearning for this kind of amiability, and, alone in her circle, she had an excuse for rejecting it, since parents, priests, and busybodies must hesitate to badger into marriage a girl with a celebrated preempted heart. Probably she was always grateful to Rezanov, clearly as she may have come to understand his character, for providing her with the precious luxury of choice in the way she was to live her life. Because of him, she could escape the annual pregnancies, the lost teeth, the conglomerations of fat, and move freely and lightly in her own world of the buoyant spirit.

In November, 1831, Douglas wrote excitedly to Professor Hooker, "I had two days since a letter from Baron Wrangel, Governor of the Russian possessions in America and the Aleutian Islands, full of compliments, offering all assistance, backed by *Imperial favour* from the Court."

It had happened. The expedition conceived in London four years earlier by Douglas, Captain Fyodor Lütke and his botanist Karl Mertens, was a potential reality. The captain had been as good as his word; on his return to Saint Petersburg he had set the wheels in motion which resulted in the invitation, delivered to Douglas in Monterey, to cross Siberia as a guest of the Russian government. All Douglas had to do was get to Sitka, headquarters of the Russian-American Fur Company in Alaska. An imperial warship would then deliver him to Okhotsk on the Siberian mainland, from which point he could start his march across the continent. The month of May, 1832, was suggested for his time of arrival at Sitka.

"What a glorious prospect!" he wrote again in this same vein. "Not only the plants, but a series of observations [made by] the same individual on both Continents, with the same instruments, under similar circumstances and in corresponding latitudes!"

He was in a fever of impatience to get started. He had done all the botanizing possible in California for the season. For months now "every bit of herbage [had been] dried to a cinder" and would not get green again until the rains. He realized that he had hardly scratched the surface of the country's floral treasure box ("it would require at least three years to do any thing like good in california") but he felt he'd made a respectable showing, collecting "five hundred species a little more or less" of which he optimistically hoped "there may be about three hundred and forty new." *

It was still too early to think about traveling to icebound Sitka, but

* In the final count, botanists credit him with sixty-eight "new" species in California.

his thoughts turned to Fort Vancouver, some eight hundred miles closer to the Russian trading post than Monterey, so he began looking for "a vessel [to] the Columbia, in which . . . to renew my labours in the North."

His watch for such a vessel was terminated by an unexpected development. Before November was out, a genial botanist and Irishman called Thomas Coulter shipped into Monterey. The men were surprised to meet; neither thought there was another plant collector within a thousand miles. But there was no hostility, at least on Douglas's part. "Dr. Coulter . . . has arrived here," he informed Hooker, adding with a burst of genuine delight, "And I do assure you from my heart it is a *terrible pleasure* to me to find a good man who can speak of Plants."

Coulter not only spoke of plants, he talked about an expedition to the Colorado and Gila rivers to collect them, especially cacti and related succulents. Douglas was intrigued. The proposed trip offered an opportunity to carry out his original instructions from the Horticultural Society in which he had been thwarted earlier in the year by travel restrictions. The restrictions seemed no longer to apply—at least Coulter wasn't hampered by them as he bustled around town engaging guides and muleteers, and Douglas was delighted when he was asked to join the party. He was even willing to risk the blazing sun and formidable heat of the southwest desert, since it was still midwinter.

But there was to be a hitch in departure. "Abominable" California politics, which had rumbled but not erupted for most of a year, suddenly burst forth with martial activity. The Paisanos of Los Angeles, growing ever bolder in their disregard for the central government, threatened outright confiscation of church lands in the south unless the padres accepted immediately a modified form of secularization. The Franciscans appealed to Monterey for help, and Governor Manuel Victoria realized that he must act with speed if anything was to be left of his authority.

A resolute soldier (he had risen from the ranks during the Mexican revolution), he mustered a detachment of some twenty men and headed south, reaching the threshold of Los Angeles in less than eight days of hard riding. The Paisanos were almost taken by surprise but recovered in time to meet him with some two hundred horsemen. Victoria was undismayed by the odds; he had survived enough *opéra-bouffe* battles to know that numbers could dissipate quickly in the face of forceful action.

The encounter took place at the Cuhuenga Pass, a defensive strong point in the hills ringing Los Angeles and appropriately close to the location of the present Hollywood Bowl. After some desultory shooting, Victoria led his men, waving guns and shouting at the top of their voices, in a charge. As he had anticipated, the Paisano forces broke and fled. The field was his, with total casualties of two men wounded, one on each side.

Unfortunately, most unfortunately and disastrously, the wounded man on the governor's side was himself. He had been hit in the thigh by a stray bullet and was carried to Mission Gabriel in considerable pain for medical attention from the padres.

Word spread fast and, by nightfall, the Paisanos had regrouped and were besieging the mission, demanding Victoria's surrender. A parley was held. Suddenly and unaccountably the governor gave up. It has been rumored that he was hit elsewhere than in the thigh, but, whatever the region, all fight left him and he agreed to yield power to the southern rebels in return for safe passage back to Mexico.

When news of these developments reached Monterey it caused shocked disbelief. The governor's capitulation was thought to be a lie concocted by the Paisanos, since obviously he had no authority to turn the country over to anyone. Captain Zamorano, second in command to Victoria, refused to recognize the Paisanos' proclamation of a new regime, and called for volunteers to protect the capital against forces that might march north to seize it. He persuaded Hartnell to organize the foreign residents into a special unit called *La Compañía Extranjera*, and, in due course, Douglas and Coulter were pressed into service as reluctant draftees.

Duty in the Company of Foreigners was not taxing, requiring only occasional nights of sentry alert against an enemy that was clearly not going to materialize, but while the crisis lasted no one was allowed to leave Monterey, which meant an indefinite postponement of the desert botanizing expedition. It wasn't until the end of March that the all clear sounded and by then it was too late for Douglas. He saw Coulter off for the Colorado River; then he kept watch on the harbor, hoping for a ship that might move him up the first leg of what promised to be a zigzag passage to Sitka.

An average of two ships called at Monterey each month, but few could offer Douglas an assist to the north. Most were droghers, the slow-moving freight boats from Boston and Europe which, as Richard Henry Dana was to describe so graphically in *Two Years Before the Mast*, shuttled between California ports filling their holds with hides and tallow. An annual ship from Fort Vancouver put in an appearance, importing lumber and taking out pickled beef and furs; occasionally a Russian ship came on a diplomatic mission. The regular supply ship for Sitka, however, always stayed clear of Monterey. It docked in Bodega Bay, close to Fort Ross, where such furs and foodstuffs as the inept garrison had managed to collect were loaded aboard with much secrecy.

Douglas was unlucky that spring of 1832 in that no ship from Hudson's Bay or from Imperial Russia put into Monterey. By mid-April he was resigned to the fact that his trip across Siberia was off for that year. He wasn't too concerned about it. Baron von Wrangel would be well aware of the difficulties of transportation, and, since events progressed in seasonal cycles in those remote regions, Douglas had no doubt that his welcome and the promised travel arrangements would hold equally good for 1833.

He could, of course, have returned to Fort Ross and, armed with the baron's letter, have obtained passage to Sitka by the annual supply ship that fall, thus insuring a May start the following season. But pride kept him from this course because it had not been suggested in the original proposal. Besides, what he had observed overnight of one Russian outpost hardly predisposed him to pass an entire winter in another. So he determined on a return to the Columbia, and from there hoped to find passage on a ship trading up the Alaskan coast, unless—and he must already have begun thinking about the possibility in Monterey—unless he could somehow devise a new and dramatic approach to the Russian stronghold.

Meanwhile in the Hartnell home he worked on his collection of seeds and specimens under the admiring eyes of Juan de la Guerra y Noriega, the youngest son of Don José, the patriarch of Santa Barbara. This unfortunate young man of twenty-one was the victim of a little learning, a dangerous thing in bucolic California. At the suggestion of Hartnell, then about to marry Juan's sister, Juan had been sent to school in England, and returned to face the taunts and jeers of his contemporaries. The illiterate young caballeros of Santa Barbara, unable to get over the fact that Juan could read and write Latin, made life so miserable for him that he moved to Monterey. He found stimulation in Douglas's company; in fact, he became so fascinated by the science of growing things that, encouraged by Douglas (who Anglicized his name to John Noriega), he began compiling California's first work on farming, calling it a "Dictionary of Agriculture."

About this time Hartnell, his fortunes at a low ebb, conceived the idea of a boarding school for the sons of the province's leading families. The curriculum was to be built around a unique faculty. Hartnell, a celebrated linguist, would teach languages; Douglas agreed to take charge of the natural sciences as long as he was available; John Noriega, with a gift for figures, would be responsible for mathematics, while Father Patrick Short, an Irish priest run out of Hawaii by the bigotry of Protestant missionaries, would conduct religious instruction. The site of the school was to be the Hartnell ranch, near the marshy area known as *Las Salinas*.

Today in the city of Salinas, California, there flourishes Hartnell Junior College, an institution "named in memory of William E. P. Hartnell, an early educator." There are two campuses sprawling over a hundred and ninety acres, one an agricultural experimental farm within a few miles of Hartnell's old ranch. The total value of the buildings and land is put at over three million dollars; the library houses fifty thousand books and the college enjoys an enrollment of more than three thousand students.

William Edward Petty Hartnell would smile at these figures. He calculated that he needed only five students for his school to pay its way. It failed because in all California there weren't five sets of parents who considered education for their sons a worthwhile investment.

In spite of political upheavals, military involvements, teaching plans, and looking out for ships, Douglas continued to botanize. When spring rains greened the land again he became a familiar figure roaming the forests and meadows around Monterey. Alfred Robinson, a cargo agent for a Boston shipping firm and a resident of the town has left us, in his *Life in California*, an account of Douglas at this period which, though striking, displays some ignorance of botanizing objectives:

> "... he'd frequently go off, attended by his little dog, and with rifle in hand search the wildest thicket in hope of meeting a bear; yet the sight of a bullock grazing in an open field was more dreadful than all the terrors of the forest. He once told me that this was his only fear, little thinking what a fate was in reserve for him."

We suspect that "friend Robinson," as Douglas called him, was strongly influenced by hindsight in remembering Douglas's "only fear," just as another historian might effectively "remember" that Goliath was terrified of slingshots or Pyrrhus, king of Epirus, of loose roofing tiles. Knowing how it all came out is a great asset in the dire-prediction business.

During the early months of 1832 Douglas gathered about one hundred fifty plant species, bringing his California total to some six hundred fifty. A feature of the collection was the high proportion of annuals, plants which, acclimatizing to the short season of rains, completed their growth cycles within a few months. Annuals were still something of a novelty in Europe. English flower gardens especially were still geared to perennials, to the iris, phlox, lilies, and primroses that Douglas had sent back in such quantities from the Columbia. But the flowers developed from his California seeds were so gay and novel as to start a whole new vogue. New methods had to be devised for growing and displaying them, resulting in a technique called "carpet bedding," or the close planting of like annuals for the massed color effect of their blooms.

There was to come a time when the carpet-bedding concept was to be abused, when hearts and cloverleaves, squares, circles, and even block letters spelling MOTHER were carved out in lawns and crammed with gaudy

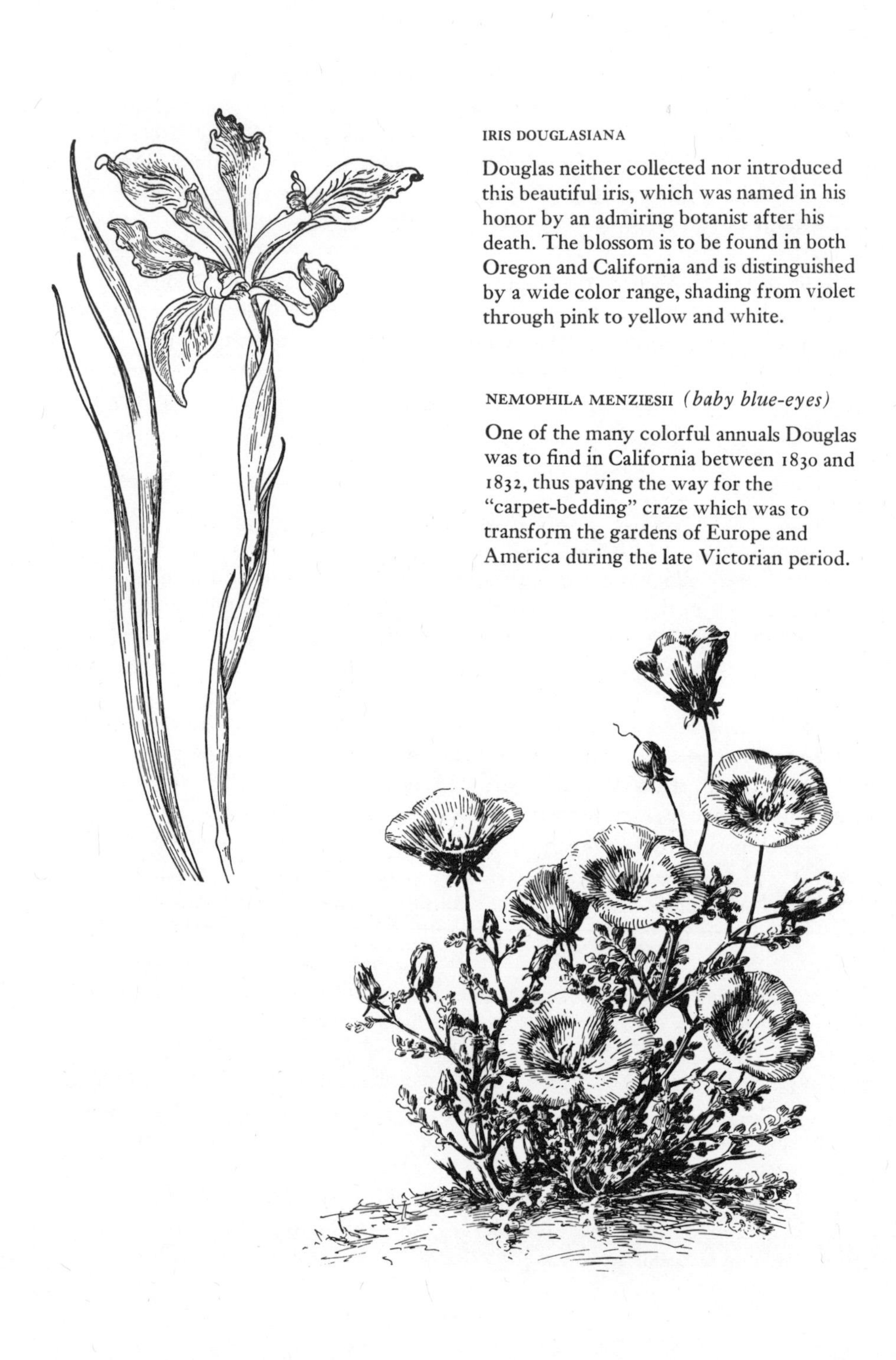

IRIS DOUGLASIANA

Douglas neither collected nor introduced this beautiful iris, which was named in his honor by an admiring botanist after his death. The blossom is to be found in both Oregon and California and is distinguished by a wide color range, shading from violet through pink to yellow and white.

NEMOPHILA MENZIESII *(baby blue-eyes)*

One of the many colorful annuals Douglas was to find in California between 1830 and 1832, thus paving the way for the "carpet-bedding" craze which was to transform the gardens of Europe and America during the late Victorian period.

blossoms. The craze for these exercises in bad taste peaked in the late Victorian period, though it carried over into the present century.

But in the beginning, when California poppies, for instance, were planted as gay sheets on manor house terraces, the effect was alive and exciting. Douglas would certainly have approved; the patches of color would have reminded him, however faintly, of the enormous poppy fields covering the foothills of Santa Barbara, stretching as far as the eye could reach, called by the padres the Altar Cloth of San Pascual, and by sailors at sea, viewing the sight from as much as forty miles out, *La Tierra del Fuego*, the land on fire.

Douglas was still waiting for his ship when Thomas Coulter returned from the Colorado desert in July, his mules loaded with floral booty. Douglas must have had his regrets at a missed opportunity as he looked through the choice collection, but he felt happier when Coulter generously offered to share duplicates with him, and also when he learned of the readings Coulter's thermometer had registered * in the Yuma territory.

Finally in August a craft of frightening dimensions arrived in harbor. It was a sloop of forty-six tons, less than half the size of the *Santa María*, smaller even than the *Niña* and *Pinta*. It was bound for neither the Columbia River nor Alaska, but for the Hawaiian Islands two thousand miles to the southwest. Even so Douglas, unable to wait any longer and gambling that his chances of shipping north were better in Honolulu than Monterey, engaged passage on the diminutive vessel. He had much to do to carry out the ambitious plan now forming in his mind, a plan involving an unprecedented cross-country journey and the virtual guarantee of a hero's welcome when he reached Sitka.

* As high as 140° F.; unquestionably this is another instance of "instrument error" since it exceeds the world's record high, 136° F. in Tripoli, North Africa, 1922. Yuma's high, 127° F. in 1905 is, in all conscience, hot enough.

Douglas's Travels
in the UPPER (Canadian)
NORTHWEST
N
CANADA
ALASKA
Sitka
50°40'
BABINE L.
Fort Simpson
SIMPSON R. (SKEENA R.)
L. STUART
Fort St. James
STUART R.
NECHAKO R.
Fort George
RED ROCK RAPIDS
FRASER R.
COAST RANGE
NEW CALEDONIA
PACIFIC OCEAN
Alexandria
COLUMBIA
VANCOUVER ISLAND
FRASER R.
THOMPSON R.
Fort Kamloops
OKANAGAN L.
R.
CANADA
U.S.
Fort Okanagan
OKANAGAN
0 Miles 200
PUGET SOUND
COLUMBIA R.
Fort Nisqually

11

JOURNEY TO NOWHERE

It was an age of heroes. Though it was to be eight years before Thomas Carlyle delivered his celebrated lectures on "Heroes and Hero Worship," the philosophical basis for them had been accumulating for half a century. The Romantic Movement, which produced in England such soaring voices as those of Keats, Shelley, Byron, and Sir Walter Scott, found heroic counterparts among men of action in such as Nelson, Wellington, and, in Douglas's time, the explorers of Polar sea.

"We all love great men," Carlyle tells us. "Love, venerate and bow down submissive before great men." Douglas certainly did his share of genuflecting. In both journals and letters he rarely mentions without suitable awe and respect the names of Captains Franklin, Parry, Ross, Beechey, and other institutionalized heroes of the northern icecaps, and his admiration for the men of action with whom he was on closer terms, the Edward Sabines and Alexander Roderick McLeods, was unrestrained. As we have seen, he could even summon up enthusiasm for relatives, comrades, and even mere acquaintances of great men. For years it seemed that no reflection from a distant star was too dim to awaken Douglas's esteem.

Then a change took place; at some point he grew tired of obeisance and decided that he had it within him to become a hero himself. We don't know precisely when this conviction developed, though we can pin it down within certain limits. It must have happened during his stay in California, since he went there solely in the interests of science but left it in search of much more. He could hardly have been filled with a sense of destiny in northern California when he failed so dismally to mount an expedition into the Umpqua country, so the time is narrowed to after his return to Monterey.

Perhaps it was watching Thomas Coulter organize his foray into the Colorado desert that first stirred Douglas to his own possibilities. Coulter, after all, chubby, peppery, and unbuttoned, was cut to no heroic pattern, and as Douglas watched him wheedle travel permits from government officials and persuade reluctant *mestizo* guides to accompany him into the furnace of the desert, he must have concluded that if Coulter could do it, so could he. All that was needed was determination and perseverance and the refusal to take no for an answer. The man who didn't doubt himself couldn't fail; or, as Carlyle was one day to put it, "A Hero is a Hero at all points; in the soul and thought of him first of all."

Douglas's heroism of soul and thought was to be put to a searching test within a few hours of his arrival at Honolulu (then called Fairhaven), the tiny sloop having completed an uneventful voyage across the Pacific despite Douglas's complaint that it was frequently lost between the waves. Lodging at the British Consulate, Douglas received mail which informed him of startling developments at the Horticultural Society in London. An administrative upheaval, following recriminations, charges, and countercharges, had forced Joseph Sabine from office as honorary secretary. Douglas's response to the news was immediate and rash. He wrote to the council announcing his own resignation in protest, relinquishing both his membership in what he was thereafter to call the "Beastly Club" and his post as the Society's collector.

From any reasoned viewpoint he overreacted to the situation. He had never been bound by close ties of either loyalty or affection to Sabine; in fact, he had frequently chafed under the Secretary's superior attitudes from the time of his first arrival in Chiswick as an untried gardener. Whatever gratitude he may have come to feel later for such things as Sabine's encouraging letters while he was on the Columbia or his efforts to introduce him into society on his return to London, had been more than repaid by a host of plants named in his honor.

Besides, Douglas had been away from England for almost three years when he learned the news, so he could hardly expect to know all the ins and outs of a complicated situation.* Under the circumstances, a letter of

* Sabine seems to have been guilty mainly of bad judgment and neglect of duties. He allowed arrears in membership dues to pile up, straining the Society's credit, which he

sympathy to Sabine together with suitable expressions of continued regard would have satisfied the amenities. The letter he *did* write was received in London as a gratuitous insult, an arrogant gesture of ingratitude from an employee of ten years' standing who owed his not inconsiderable botanical reputation to the organization he now saw fit to malign.

Douglas could have had no doubts about the effect of his letter. He knew it must lead to his financial and professional bankruptcy. Once word was out that the Horticultural Society was no longer his employer and guarantor, his income would end, his credit would end, and the hospitality he had enjoyed on the Columbia, in California, and even now in the home of the British Consul in Honolulu, would no longer be possible. There would come a time when even complimentary transportation on British ships, now guaranteed by his credentials from the Colonial Office, would be denied him, raising the specter of being stranded at the ends of the earth.

The question occurs and recurs; why did he do it? How could he become so indignant over the now stale misfortunes of an indifferent friend half a world away that he was willing to destroy his own future for him?

He provides no answer in a letter written to Professor Hooker shortly after the event, but he does provide a clue in that he assumes that Hooker understands that resignation from the Society was the only honorable course open to him. It was the feudal position of *noblesse oblige*, the kind of unquestioning loyalty one knight owed another irrespective of the rights or wrongs of a situation. In short, it was the posture of a hero.

But there is at least the possibility of a second motive for the ill-advised letter: fear of himself. Unquestionably the projected trip to Sitka now fully occupied Douglas's mind, but, despite his conscious determination to succeed at all costs, he may have been troubled by a small voice that whispered he would never have the courage to go through with it. The resignation from the Society may well have been designed to close the last avenues of retreat upon himself. It smashed the lifeboat, kicked away the ladder, and burned the last bridge. It made success mandatory. Douglas had before him eighteen months, perhaps two years, before word leaked through official channels that he was a floater. By then he must either be dead or halfway across Siberia on his way to England.

After all, if he returned to London at the end of an unparalleled scientific expedition to be hailed as the first traveler to have walked around

---

strained further by staging expensive fetes and exhibitions on borrowed money. He resigned in the face of a committee investigation which questioned his competence but not his integrity.

the world (barring only the Atlantic Ocean), he was hardly likely to need the Horticultural Society ever again.

The mood of destiny still clung to Douglas when he arrived at Fort Vancouver aboard a Hudson's Bay Company ship in mid-October, 1832. His man, William Johnson, reassigned to his service, was delighted to have him back, as were others, but by no means all, among the fort's personnel. Chief Trader Archibald McDonald, for one, had an adverse reaction. He found Douglas's new self-assertive manner a strain. He confided in a friend that, to hear the returned traveler boast about his hunting exploits in California, "bears, bulls and tigers had cause to rue the day they went there."

Though he had been absent for only two years, Douglas found that civilization had encroached with giant strides upon the fort. Many of the great Douglas firs had been cut to feed the new saw mills, while others had been simply felled and burned to clear the land for agriculture. More than two hundred acres were under cultivation, providing stands of wheat, corn, and vegetables and a famous apple orchard.

With the price of beaver skins falling steadily on European markets, Dr. McLoughlin had built up a brisk international trade in other products. The main traffic was with the Hawaiian Islands, where Fort Vancouver's timber, smoked salmon, and flour were traded for coffee, sugar, molasses, and rice. Two Hudson's Bay Company ships engaged in this commerce, and a third one, a seventy-ton schooner, was currently under construction on the river's bank. California ports and the Russian posts in Alaska also were visited by Columbia River ships, but not on a regular basis.

Douglas found the changed landscape depressing and complained of "too much civilization." The deer which once picturesquely dotted the meadows around the fort were gone, hunted to extermination in order to protect the crops. Even Indians were rarely to be seen plying the river or trooping into the fort to trade foreign trinkets.

There were, in fact, few Indians to be encountered anywhere on the Columbia, up or down the river. Epidemics of the strange "intermittent" fever had decimated the tribes during the past few years, though "decimated" is an understatement of the actual death totals, estimated by Dr. McLoughlin to be as high as 75 percent of the population. He imported two physicians from Scotland to help him stem the tide of fatalities, but no effective treatment ever seems to have been developed.

Two years before, Douglas had written Professor Hooker about the situation. "Villages, which had afforded from one to two hundred effective warriors, are totally gone; not a soul remains. The houses are empty and flocks of famished dogs are howling about, while the dead bodies lie strewn in every direction on the sands of the river."

As a pathetic tribute to the trust placed in Dr. McLoughlin, whole families would sometimes crawl close to the fort to die, confident that he would arrange for decent burial. Too often, however, he was unable to carry out the trust; his hands were too full with his own sick and dying patients within the fort. Both white and red inhabitants came down with the chills, shakes, and incapacitating fevers, but, as with other imported diseases, the Europeans tended to recover while the Indians died.

Medical opinion is cautious about putting a specific name on the "intermittent" fever. The symptoms were certainly malarial, but a confusing factor was that mosquitoes had swarmed since time immemorial in the Columbia territory without causing an affliction worse than itching skin. However, there seems at least the possibility that the malaria-carrying anopheles were newcomers to the region, possibly arriving aboard ships from infested South America, and when once acclimatized, they began their deadly havoc.

In February, 1833, in a winter so cold that six inches of ice formed on the Columbia, Douglas restlessly embarked with Johnson, Billy, and a complement of *voyageurs* to visit Fort Nisqually, a new Hudson's Bay Company trading post on Puget Sound. Much of the route lay up the Cowlitz and Chehalis rivers, regions through which Douglas had returned from his one completely joyful visit to Cockqua's lodge. But if he was romantically stirred, he gave no indication as he methodically set up his surveying instruments and searched the heavens for bearings.

Certainly he made no attempt to push on down the Chehalis to its mouth, though at times he was no more than a day's journey away. Either he found the thought of return unbearably painful or, considering the wide devastation of the fever, he may have presumed that Cockqua and his tribe were wiped out.

He could have been surprised at his old friend's resilience. Some twenty years later a visitor to the Chehalis found the villages of the estuary still intact, though vastly reduced in populations. He remembered especially a dignified elder by the name of Ka-kow-an who, speaking excellent English, inquired after the health of King George's chiefs. Ka-kow-an, Cockqua; it's close enough. He was such stuff as legends are made on.

Back at Fort Vancouver, Douglas began preparations for his Great Expedition, though he was concerned by a new development in the recurring troubles he had been having with his vision. For some five years he'd been experiencing symptoms that alarmed him because of their perplexing and contradictory nature. At times, though his distance perception was excellent, he was unable to focus his eyes to read; at other periods, he had trouble distinguishing objects at long range, though his ability to shoot a rifle accurately was unimpaired. Now, as the most recent cause for anxiety, his left eye was gaining in clarity and sharpness while his right eye had faded into a blackness "as dark as midnight."

Actually, though Douglas may have found it hard to believe, his prognosis was good. For, even at this distance, it is reasonable to surmise that he was afflicted by anisometropia, a condition in which the refractive powers of the eyes differ one from the other. Nature, left to itself, has a way of solving the conflict by suppressing the use of one eye (Douglas's right eye) thus permitting unimpeded and satisfactory functioning of the other. And so it seems to have worked out in Douglas's case; at least we hear no more about defective vision.

Fortunately the urgency of Douglas's preparations didn't give him much time to brood about his health. He was due to leave Fort Vancouver with the eastbound Hudson's Bay express in March, ostensibly to botanize in the Kamloops territory north of Fort Okanagan, but in reality to begin the first leg of his march to Sitka.

Because of the devious negotiations lying ahead of him to obtain men and supplies, Douglas dared not reveal the true purpose of his trip to anyone lest word get back to Dr. McLoughlin. But his need for secrecy put a strange light on some of the purchases he must have made at the Company storeroom; for instance, winter clothing for a summer's botanizing expedition, and quantities of powder and shot more appropriate for an extended expedition than a few weeks' jaunt through the pine woods.

Sitting at the gentlemen's table in the place of honor to Dr. McLoughlin's right must have been an ordeal during those final days. There can be no doubt about Douglas's regard for the doctor; he had enjoyed too many courtesies and accepted too many kindnesses and favors from him since he first arrived on the Columbia to feel anything but gratitude and respect. It was a poor return that, within weeks, he was to be maneuvering in his own interests behind his host's back.

He must have been tempted to come out into the open, to confide his plans to the doctor and ask for cooperation. But he knew that he couldn't. For the chief factor was, after all, the Hudson's Bay Company's supreme commander in the west, and a refusal from him would end the prospect of help from any other quarter.

Even so, when the day for departure came, when Douglas took his place in the express boat, he must have coped with strong emotions as he watched the giant figure of the doctor on the wharf, dressed in his perpetual black, his white hair flowing, as, with gold-headed cane raised in one hand and tall beaver hat in the other, he released the expedition with a cry of "Godspeed!" For Douglas never expected to see either Dr. McLoughlin or Fort Vancouver again.

The first part of the trip, the laborious five-hundred-mile pull against the current, was an old story to Douglas, but remembered landmarks and recalled memories helped dispel the tedium of what he expected to be his last voyage on the great river. At Fort Okanagan he left the express and, with Johnson and Billy, instruments and baggage, joined a horseback brigade headed into "the interior," that enormous, largely unmapped and unexplored territory to the north called New Caledonia.

The route followed the Okanagan River to the lake, then cut across country to the Thompson River and Fort Kamloops, now in charge of Douglas's old friend from Walla Walla, Chief Trader Samuel Black. For most of the way it was a land of low rainfall, sweeping winds, and sparse vegetation, a continuation of the great arid region to the lee of the Cascade Mountains. But as the Thompson River came into view, so did trees, at first in groups and copses, but later in handsome forests spreading over the hillsides.

Douglas did little botanizing during the two-hundred-mile trip, which took several days. Since breaking with the Horticultural Society, he had been collecting exclusively for Professor Hooker, a specialist in mosses and lichens, and there were few of either to be had in that largely dry terrain. But he never failed to take observations of the sun and stars every twelve hours and often worked up additional data in the form of botanical notes and rough sketches of geographic features, clearly experimenting with scientific techniques that might be useful in Siberia.

Samuel Black had obviously benefited from the change of scene. His "zeal," a quality on which he had once prided himself but which seemed at times to have been baked out of him in the adobe oven of Walla Walla, was back in force. Douglas felt it in the vigorous welcoming handshake whose warmth he tried to gauge as if he held a thermometer in his hand. He was staking much on that warmth, on that special regard that Black had once shown for him. It was vital to the success of his plans.

There was a reunion with Angelique Cameron and the children. Douglas must have brought presents, hair ribbons for the girls, pocket knives and English fishhooks and lines for the boys. In the ensuing excitement Douglas would once more be aware of the privileged position he enjoyed in that household; Black always kept his family out of sight when his Company colleagues came upriver, even though most of them were also fathers of mixed broods.

There was a brigade stop of several days at Fort Kamloops in order to rest men and horses for the grueling trip on to the ultimate outpost of the Hudson's Bay Company empire at Fort Saint James. Douglas would take his time. The first evening would be spent in catching up on news; or, perhaps more accurately, Douglas would catch Black up on his triumphs in England and travels in California while the men sat late over brandy. If Black volunteered information, it was probably to the effect that beaver-skin collections had been good at Kamloops because local conditions were hard enough to force the Indians to exert themselves.

Finally, a night or two before departure, Douglas broached his subject. He would explain the unique nature of his projected trip across Siberia, its unprecedented scientific possibilities, and produce Baron von Wrangel's letter to show what arrangements had been made. The only gap still unprovided for in the journey was the five-hundred-mile stretch between Fort Saint James and Sitka, a span never crossed by a traveler before, and he hoped Black would help him bridge it. He would need guides, men, and supplies, and was depending on Black at Kamloops and Chief Factor Peter Warren Dease at Fort Saint James to supply them.

Black must have listened with mild incredulity to this proposal. He would ask immediately if the trip had Dr. McLoughlin's authorization, and, if so, why hadn't he been informed? When he learned the truth, he may have been more puzzled than anything else. How could Douglas expect *him*, acting without permission, to commit the Company's men and materiel to a scheme that had nothing to do with the fur trade?

Douglas had an answer to that. He had come to Black because he knew him to be a man of decisive action, impatient of red tape. He had proved himself capable of unorthodox tactics when, as a young Northwester, he had hounded rivals out of the Athabaska territory. He hadn't waited for authorization then but had gone to it with a carefree heroic spirit. Surely he would support Douglas now in another kind of large enterprise that promised a glorious ending.

But Douglas was to find that appeals to glory were lost on Black these days. His adventurous youth was far behind. He was a conservative now, interested in the Company's profits, not in heroic exploits. Besides, he argued, Douglas's proposal made no sense. If he wanted to get to Sitka, why not return to Fort Vancouver and wait for a ship bound for Sitka? Why the necessity to go overland?

It must have been difficult for Douglas to remain patient. He had deserved better of Black than this, especially since he suspected that the obtuseness was being assumed as a protective device.

Perhaps it was Black's ring, dedicated "To the most worthy of the worthy Northwesters," which suggested the next avenue of approach. He, of all people, must understand the heroic drive. When the great Sir Alexander Mackenzie made his historic dash across the continent to the Pacific, nobody asked him why he didn't take a ship. And nobody asked foolish questions of David Thompson or Simon Fraser when they rode their rivers to the ocean. All Douglas wanted was the same consideration. He wanted to reach Sitka overland because, as he had stated, it had never been done before.

We don't know how the chief trader answered that argument, but we do know, graphically, how Douglas answered *him*. We allow the historian Hubert Howe Bancroft to take over the account:

> While enjoying the lonely hospitality of his brother Scot, and discussing the affairs of the company, Douglas, who was more fiery than politic, exclaimed: "The Hudson's Bay Company is simply a mercenary corporation; there is not an officer in it with a soul above a beaver-skin."
> Black was up in arms in a moment. He informed his guest that he was a sneaking reprobate, and challenged him to fight. As it was then dark the duel was postponed until next day. Bright and early in the morning Black tapped at the pierced parchment which served as a window to the guest-chamber, and cried out, "Misther Dooglas! are ye ready?" But the man of flowers declined the winning invitation, and saved his life.

It must have been an agonizing experience for Douglas. After so much high-flown talk he was failing to live up to his own image of heroic conduct. But his string wasn't quite played out yet, and until that happened he was unwilling to risk wounds or death, the inevitable outcome of the duel since he had no intention of firing at Black.

So he waited out the humiliation in silence. His last slim hope now was to get on to Fort Saint James. He had never met Chief Factor Peter Warren Dease, commander of that post and brother of the tea drinker, but he had heard a great deal about him, most of it encouraging. He seemed cut from the same cloth as Captain John Franklin, Edward Sabine, and Alexander Roderick McLeod, and was hopefully of their spirit.

Peter Warren Dease was, in fact, an extraordinary man. He had never thought of himself as anything but a fur trader until the Hudson's Bay Company's Governor Simpson lent him to the second Franklin Polar Expedition as a commissariat expediter. He did an excellent job of provisioning that gallant but rather inept band of heroes through the winter of 1825–26, then returned to the fur trade apparently immune to exploration fever. Nevertheless, the virus had entered his system, and, though kept under control for more than a decade, it could not be suppressed forever. In 1837 he was to take a leave of absence from the Company and head north for polar shores. Methodically he linked up the scattered discoveries of Franklin, Beechey, Parry, Ross, and others, filling in the blanks with his own surveys until finally the entire continental coast, off which lay the fabled Northwest Passage, stood charted and defined. What the British Admiralty had failed to achieve in a dozen years of perilous voyages and overland expeditions, Peter Warren Dease accomplished in just two seasons with a party of fourteen men and two boys. He was rewarded by a trip to London to see the Queen and meet the Hudson's Bay Company's directors, though he was to enjoy more enduring honors in the Arctic river, bay, and strait that bear his name.

When Douglas met him, Chief Factor Dease was in his in-between or suppressed-fever stage, his first brush with glory behind him and his great exploring feats still in the future. He was forty-five years of age, and probably still fitted the description Governor Simpson had once given of him, "of a strong robust habit of body, possessing much firmness of mind joined to a great suavity of manners."

Douglas probably saw the suavity first as his host gave him a chance to rest up after the rough five-hundred-mile journey from Kamloops. But the day came when Dease, probably prewarned by a letter from Samuel Black, informed Douglas agreeably but firmly of the ways he could, or could not, be of assistance in his travel plans.

A special brigade organized to take Douglas to Sitka or, rather, the mainland opposite the island on which Sitka was situated, was out of the question. And not only for policy reasons. Even if Douglas had arrived fully authorized and equipped for the journey, Dease would have tried to discourage him. The route lay through unknown, trackless country peopled by Indians so primitive that they subsisted, at least in part, on a diet of prisoners captured in their ceaseless warfare.

On the positive side, Dease suggested an alternative method by which Douglas might reach Sitka from Fort Saint James, since that approach seemed to be of such importance to him: A party of hunters was shortly to travel down the Skeena River to investigate trapping potentials, and Douglas was welcome to join them. If he decided to go, Dease would arrange to have him escorted up the coast to Fort Simpson, the Hudson's Bay Company's most northerly establishment in New Caledonia. From there he might be able to cover the final three hundred miles to Sitka by a coastal trading vessel, if one ever called, which it rarely did.

Douglas climbed the hills behind the fort to think matters over. Though June, it was still cold, with patches of snow lying beneath the trees. Looking down on the rough-slabbed, clay-calked buildings, he had no difficulty in imagining the howling gales of winter when temperatures could drop to fifty-five degrees below zero. And in other respects it was an inhospitable land. Food was always a problem; the lake supported a meager fish population, and it was said that a deer hadn't been seen in a hundred years.

Still unsure as to his future course, Douglas stayed on at the fort for several days, botanizing listlessly. As far as possible he avoided the dark-visaged, almost black Indians of the local tribal group. They were an enigmatic people called "Carriers" by the traders because of their strange necromantic beliefs: widows were forced to carry the charred bones of husbands snatched from funeral pyres until the shaman declared the souls at peace. Douglas, prejudiced from the start against the Carriers because of their complexions, was utterly revolted when he heard of this custom; so revolted, perhaps, that he was finally turned against the country and everything in it. All that's certain is that when the hunting party left for the coast Douglas was not with it. Instead, he was in a canoe with Johnson, Billy, and his belongings, paddling down the Stuart and Nechaco rivers headed for Fort George on the first leg of the long journey back to Fort Vancouver.

He had traveled eleven hundred and fifty miles buoyed by hopes and a dream, only to be defeated by realities. It was going to be difficult to readjust, to pare his ambitions to more obtainable ends. And perhaps for Douglas, who from childhood on had never been able to strike a compromise between what was practical and what he wanted, the readjustment was impossible.

Fort George (now Prince George), situated at the head of the Fraser River, is entitled to a certain mild fame as the take-off point for celebrated voyages. It was from Fort George that Alexander Mackenzie started his epic journey south and west, and it was the power of the current combined with the tranquillity of the flow that led him to believe that the river was the Columbia. Simon Fraser also started at Fort George some fifteen years later also thinking he was on the Columbia since Mackenzie had not explored far enough to learn anything to the contrary. Both men soon found, however, that the placid stream off Fort George changed some twenty miles to the south into a roaring, raging maelstrom as the waters rushed through a canyon chute called in their time the Stony Islands but now more usually known as the Red Rock Rapids.

Douglas knew all about the rapids, of course. He had looked down on them during the passage upstream as he walked along the canyon rim while the boats were being portaged. He knew what vast respect the *voyageurs* had for those swirling waters, and he knew that Hudson's Bay Company standard operating procedure called for the rapids to be portaged both going up and downstream. So there was no excuse for what happened that fine morning in June, 1833, as Douglas and Johnson, having pushed off from Fort George at dawn, paddled confidently toward the gorge without alarm at feeling the accelerated current under them or hearing the crash of waters up ahead.

It is possible that Douglas had decided on this form of Russian roulette because he remembered that, although Alexander Mackenzie had cautiously set the precedent of portaging the rapids in 1793, Simon Fraser in 1808 had boldly shot them. We don't know when the idea lodged in his mind that if Fraser could do it, so could he, but the prospect of the adventure understandably revived his spirits as it challenged his drooping sense of the heroic. What he didn't remember, or didn't care about if he did, was that Fraser's brigade had consisted of three stout boats, manned by nineteen tough and experienced *voyageurs*, which gave it odds in favor of success not shared by two tenderfeet and a dog in a birchbark canoe.

No doubt Douglas consulted Johnson before committing him to peril, and no doubt he received a thumbs-up response from the old man-o'-war tar, real or imagined, before heading the craft into the foaming water that roared through the canyon of rocks.

When the canoe hit, to be "dashed to atoms," Douglas was thrown clear. He was pummeled, rolled around, and dragged downstream for an hour and forty minutes. "I passed over the cataract and gained the shore in a whirlpool below, not however by swimming, for I was rendered helpless and the waves washed me on the rocks," was the way he remembered it. Wandering back dazed along the shore, he found Johnson, mauled but alive, and later, miraculously, Billy.

But there the glad tidings ended. "On that morning at the stony islands," Douglas wrote Hooker, "I lost every article in my possession, saving an astronomical journal, book of rough notes, charts and barometrical observations, with my instruments. My botanical notes are gone, and what gives me most concern, my journal of occurrences also, as this is what can never be replaced, even by myself."

Once again, as with Dr. McLoughlin's horse, Douglas was unable to overcome his reverence for property. Struggling for life, he'd clung to all those expensive astronomical instruments bought on credit in London while sacrificing both to his own age and to posterity the products of his unique talent, his botanical contributions.

There was the humiliating return to Fort George and the necessity to wait until a fur brigade passed through, traveling south. The stopover at Fort Kamloops must have been strained. Douglas and Black probably avoided each other as much as possible; certainly neither made a move toward reconciliation. Their friendship had been totally consumed in that one flare-up, leaving no spark in the ashes.

Perhaps it didn't matter in the end, since neither man had much time left for regrets. Douglas was to be dead within a year; not too long thereafter, Samuel Black, promoted to chief factor and contemplating retirement, was killed by a distraught Indian who accused him of putting a fatal curse on a relative. The cruel hoax of the smallpox bottle had come full circle, unjustly visiting retribution on the most worthy of the worthy Northwesters.

Douglas took his time about returning to Fort Vancouver, perhaps in dread of facing up to Dr. McLoughlin. He lingered at Fort Okanagan, then on his way downriver spent several weeks at Walla Walla. His friend George Barnston had been transferred to another post, but the young clerk who had replaced him was equally solicitous of the distinguished guest who chose to share his adobe oven in July, and was delighted to accommodate him with guides, supplies, and horses when he asked to travel once again into the Blue Mountains.

There were no obstacles this time to prevent him from reaching *La Grande Ronde*. The fever that had devastated the tribes of the river had reached into the mountains too, depopulating villages, leaving the survivors hardly able to feed themselves, let alone conduct wars. The legendary valley of the *voyageurs* lay waiting for discovery; all Douglas had to do was to climb the ridges until he found it.

But he seemed unwilling to make the effort, or perhaps he had stopped believing in the valley's existence. He was content to wander through the woods and meadows he had previously explored, collecting again specimens of his great discoveries, the lupine named for Joseph Sabine, the flowering currant for Donald Munro, and the peony for Robert Brown.

It was a period in which he seemed either to live in his past or in his imagination. He had neither the wish nor the strength to grapple with his present problems. He admitted in a letter to Hooker that he was "much broken" in "body and mind" and "strength and spirits." Perhaps it's not to be wondered at that he sometimes became confused about his recent travels. "I have been in the Snowy Mountains as high as the [latitude] 60°," he claims in another letter, "over a dreary unhospitable country, where I suffered extreme hardship."

Douglas never reached latitude 60° or was even close to it. His highest northern determination was 54° 26′ 46″, at Fort Saint James. Nevertheless, latitude 60° is significant. It is well above the location of Sitka. In fact, traced to the west, it would have placed Douglas near the Russian port of Okhotsk from which he was to have begun his land journey across Siberia.

He returned to Fort Vancouver in August. Dr. McLoughlin was solicitous about his mishap on the Fraser River but made no reference to other aspects of his journey. Apparently neither Black nor Dease felt it necessary to report any difficulties or problems with their visitor.

Douglas made no attempt even to inquire about ships to Sitka. Instead he embarked on a schooner bound for the Hawaiian Islands with a stopover in San Francisco. Here he learned the distressing news that "poor John Noriega of so much promise" was dead at twenty-three, apparently of peritonitis. He was tempted to travel down to Monterey to commiserate with the Hartnells, but the uncertainties of shipping decided him against it, and so he may have missed the last great botanical opportunity of his life.

For while Douglas lingered in San Francisco, an extraordinary frontiersman called Joseph Reddeford Walker was approaching Monterey at the end of an incredible journey. Trapping in the vicinity of the Great Salt Lake, Walker and his companions got tired of the brackish water and started west in search of a cooling drink. They crossed the desert, penetrated the Sierras and kept on going until they reached the coast. They wintered in Monterey, then left California in the spring, Walker again having no difficulty finding his way through the mountains of the south by a pass that still bears his name. In the long, towering escarpment of the High Sierras from the Stanislaus River to the Tehachapi Ridge there are only two natural passes and Joseph Reddeford Walker found them both first crack—one coming in and the other going out.

Coming in through the Tioga Pass the Walker party almost tumbled into the mile-deep canyon that one day would be called the Yosemite Valley. Skirting its rim, the men descended a stream and entered a forest of enormous, redbarked trees, probably what is now known as the Merced Grove of Big Trees.

That winter, when Walker and his men tried to talk about the giants they had seen in the mountains, the residents of Monterey smiled tolerantly and pointed to their own very sizable coast redwoods. But the men knew there was a difference, and so would Douglas if he had been there and heard their descriptions. Given the enthusiasm of the men, an expedition might easily have been improvised to take Douglas back up into the Sierras to see for himself. It was an open winter and there wasn't much else to do.

It would have been a crowning triumph to have described and introduced the *Sequoia gigantea*, or whatever it would have been called if Douglas had had the honor of naming it. But who knows? Perhaps he would have ignored the excited accounts of the wild-looking frontiersmen; for the wonder and belief that had once driven him on to his great discoveries were now at low ebb.

He had given up hope of crossing Siberia or of even finding *La Grande Ronde*. How could he believe in the existence of trees massive to the point of shock, living fossils from a thousand years before the birth of Christ?

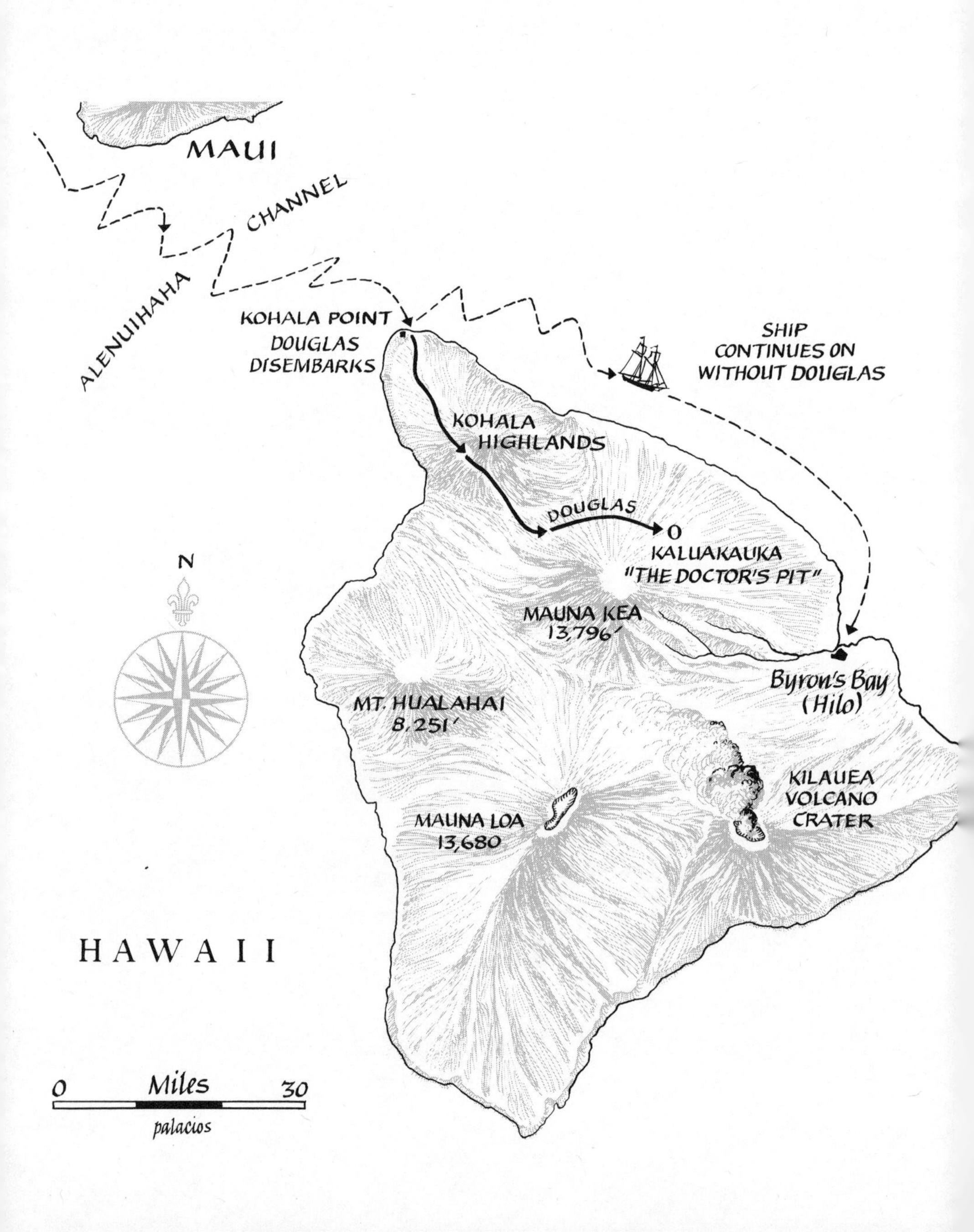
MAUI
ALENUIHAHA CHANNEL
KOHALA POINT
DOUGLAS
DISEMBARKS
SHIP
CONTINUES ON
WITHOUT DOUGLAS
KOHALA
HIGHLANDS
DOUGLAS
KALUAKAUKA
"THE DOCTOR'S PIT"
N
MAUNA KEA
13,796'
Byron's Bay
(Hilo)
MT. HUALAHAI
8,251'
KILAUEA
VOLCANO
CRATER
MAUNA LOA
13,680
HAWAII
0 Miles 30
palacios

12

DELAYED ACCOUNTING

Douglas celebrated Christmas in Honolulu at the home of the British consul, Richard Charlton. He liked his host and enjoyed the company of his wife and sister who completed the menage, but, in his current frame of mind, sitting solemnly at table with a trio of adults on a tropical island must have seemed a travesty on Christmas dinners he remembered in Glasgow, when snow outside enhanced the coziness within and the room rang with the laughter of the Hooker children.

He didn't wait around for New Year's Day, a date that in the past, had stimulated him to gloomy stocktaking, but took passage December 27 on a coastal schooner for the island of Hawaii, landing at Hilo (then called Byron's Bay) on January 2, 1834. He had exactly six months and ten days of life left to him.

His host at Hilo was the Reverend Joseph Goodrich, a pioneer American missionary. If Douglas had been given to philosophic reflections, he might have seen parallels between the Franciscan domination of California's Indians and the similar influence exerted over Hawaiian Islanders by the Protestant ministers of New England. The parallels would have been more superficial than real, however. The Spaniards imposed themselves on the native populations whereas, in Hawaii, the Americans came at the invitation of the Hawaiian ruling family to teach the islanders Christianity, a form of magic that had clearly shown itself more potent than Polynesian witchcraft.

The Reverend Goodrich, an amateur geologist and ardent mountain climber, was delighted to entertain a guest of Douglas's scientific reputation. He was so sure that the first project on the agenda would be to scale the great extinct volcano of Mauna Kea (White Mountain) that he had sixteen guides and baggage carriers to stand by for the ascent.

Mauna Kea at that time was estimated to be 18,000 feet high and was considered one of the loftiest mountains in the world. Douglas's own calculations were to deflate it to 13,851 feet, a considerable disappointment to Mr. Goodrich, who had climbed to the summit twice thinking he was outscaling Mount Ararat (16,915 feet), where Noah's Ark touched ground after the Flood. Today Mauna Kea (now officially put at 13,784 feet), though no longer counted among the giants, still qualifies as the tallest *island* mountain in the world, for whatever distinction that is.

On January 7, accompanied by the sixteen members of his expedition, most of them bent double under puzzling loads, Douglas set off for the mountain. "After passing for rather more than three miles over plain country" rich in crops of taro, bananas, sugar cane, and other fruits and vegetables, the party began to climb a wooded gulch.

"Here the scenery was truly beautiful," Douglas noted. He had started a journal again, specifically for his Hawaiian travels, intending to send it when completed to his brother John. He continued to send his botanical material to Professor Hooker, but since, in a sense, he would be in competition with Hooker as a writer, he thought it best to try to get the journal published separately.

The party pushed on up the gulch through incredibly rich vegetation. "Large timber trees were covered with creepers and species of *Tillandsia* [hanging moss] while the Tree Ferns gave a peculiar character to the whole country."

Yet beautiful as it was, lush as it was, novel as it was, Douglas must have had an uncomfortable feeling that it wasn't *his* terrain or *his* kind of material. As his friend George Barnston was to write years later: "His contributions had not been so much of a kind to increase largely the number of hot house plants, but rather of those that . . . will flourish out of doors in a temperate climate." In this work—the recognition, collection, and introduction of ornamental plants suitable for European and North American gardens—Douglas was without peer, the most confident and successful collector of his time whose record in both horticultural and botanical * discoveries is not likely to be surpassed now that the great era of plant exploration

* A measure of his botanical success lies in the esteem in which he is still held by systematic specialists. In the index of a modern compendium (Munz and Keck's *A California Flora*, 1959 edition) Douglas's name (as *Douglasii* or *Douglasiana*) is attached to eighty-six species or varieties of plants, honors bestowed on him by generations of

is over. It was only among the pampered and perishable products of lands of perpetual summer that he felt uncertain and ill at ease.

Once before, set down in another tropical paradise, he had recognized immediately that the exotics of the Juan Fernandez Islands were not for export and had wasted no time trying to collect seeds or roots. Now, however, in the balmy midwinter of Hawaii, he seemed almost desperate to find plants that might prove adaptable to London or Boston and thus give some significance to his present activities. "I observed a species of *Rubus* [raspberry] among the rocks," he noted hopefully; and again, "The ground was thickly carpeted with *Fragaria* [strawberry], some of which were in blossom, and a few of them in fruit." He seems totally to ignore the fact that berries able to fruit in Hawaii in January are not likely to do much in the botanic garden in Glasgow, even in June and up against a warm wall.

Douglas and his troupe continued to climb the mountainside. Above the gulches, the tropical vegetation was succeeded abruptly by the dry brush of the Highlands. It began to rain and Douglas was glad to accept shelter in the lodge of two American cattle hunters. He was chilled and he yearned for a cup of tea, but "unluckily, my guides all lingered behind, and those who carried my blanket and tea-kettle were the last to [appear]."

He was beginning to learn that the islanders were resistant to hustle. "These people have no thought or consideration for the morrow," he complained, "but sit down to their food, smoke and tell stories, and make themselves perfectly happy." He learned too of their enormous appetites. Only five of his sixteen men carried baggage; the rest carried the food necessary to keep the party on the road. Each man in the course of a week ate his weight in *poi,* the sticky paste made from taro roots, and meat and fruits had also to be found to fill out the meals.

When the rain stopped, the American hunters killed and dressed a couple of wild bulls as a parting gift. Douglas's porters gleefully shouldered the meat, weighing in all nearly a thousand pounds, but within a few days it was gone and a food shortage again threatened. Though he was aware that in Hawaii corpulence was considered beautiful and the most admired men and women were those who had successfully transformed themselves into butter tubs, Douglas couldn't get over the sheer bulk of food consumed. "No people in the world can cram themselves to such a degree as the Sandwich Islanders," he remarked with awe.

After two days of painfully slow group travel Douglas impatiently pushed on by himself and, at noon, January 9, stood on the summit of

botanists in appreciation of the range and scope of his collections. In comparison, Archibald Menzies, who botanized in California before Douglas, is commemorated thirty-eight times and Thomas Nuttall, who visited later, is recognized in forty-two plant names.

Mauna Kea enjoying a view that "was sublime to the greatest degree." Before making his deflationary measurements of the mountain and while still thinking that he stood at, or close to, the top of the world, Douglas was moved, as we have already quoted in part, to suitable eloquence. "Man feels himself as nothing—as if standing on the verge of another world. The deathlike stillness of the place, not an animal nor an insect to be seen—far removed from the din and bustle of the world, impresses on his mind with double force the . . . helplessness of his condition, an object of pity and compassion."

Back in Hilo, Douglas paid off his men, giving them the option of two dollars in currency or the equivalent in trade goods—printed cottons, combs, and scissors. To his dismay, most of them took cash.

His finances were inevitably becoming a matter for concern. He had tried to play fair with the Horticultural Society, calculating what it owed him in the three years of accumulated salary before his resignation, and scrupulously drawing only on that sum in Hudson's Bay Company supply rooms and other places where he had credit. He made every effort to be economical, but it was impossible for a man in his position, still believed to be a representative of a wealthy London organization, to live like a pauper, especially among missionaries who expected generous donations to their churches in return for hospitality. Besides, though he was able to keep a fairly accurate record of his own cash outlays, unseen charges over which he had no control were a continuous source of worry. For instance, at Fort Vancouver, he had reason to believe that the wages of his man Johnson had been quietly transferred from the Company's books to his account, and in Honolulu, though Consul Charlton was far too gentlemanly to suggest payment for board and lodging, Douglas suspected that chits were being sent back to his former employers, an action that would inevitably result in the cutting off of all credit once word crept back from London.

But despite financial anxieties, Douglas never hesitated to put up the money for projects he considered likely to advance science, and so, soon after his return from Mauna Kea, we find him asking the Reverend Goodrich to employ a new set of porters for a trip to the great volcano of Hawaii.

The sight of Kilauea, once described as "a purgatory of molten metals, smells and flames," proved a traumatic experience for Douglas. Only once before, at the discovery of the sugar pine, had he reacted with such excitement. Then he had been in a frenzy to measure, collect, and describe; now it was the opposite. As he arrived within sight of the crater, "a scene of all that is terrific," he was helpless to do anything but stare; "not, correctly speaking, to enjoy, but to gaze with wonder and amazement on this terrific sight, which inspired the beholder with a fearful pleasure."

What he saw was "a lake of liquid fire . . . boiling with furious agitation." As he watched, "red hot lava would dart upwards . . . with terrific grandeur, spouting to a height which, from the distance at which I stood, I calculated to be from forty to seventy feet, when[ce] it would dash violently against the black ledge, and then subside again for a few moments."

Twilight settled in. While his men cooked supper in jets of steam shooting up between fissures in the rocks, Douglas continued to stare into the boiling volcano from which emerged a noise "dreadful beyond all description."

As night fell, Douglas's tent was pitched twenty yards from the crater's edge, but he continued to sit outside while the "nearly full moon . . . shed her silvery brightness on the fiery lake." Not until morning did he realize that he had "lost a fine night for making astronomical observations."

He next descended into the crater, braving the sulphurous fumes and clinging to the partially cooled rocks at the edge of the "vast basin." He saw at close range the phenomenon known as "Pele's tresses," filaments of molten lava "twisted into a thousand different shapes, sometimes . . . like fine hair."

Pele was the vengeful goddess who, in ancient times, had been thought to dwell beneath the seething cauldron. She, along with other Hawaiian female divinities, had been replaced now by the largely male hierarchy of the Christian church, but in her time Pele had been powerful. When moved to wrath, she erupted out of her crater, and only living animals thrown into the lava flow could appease her. The Hawaiians never deliberately went in for human sacrifice but on one occasion a band of warriors approached impiously close to the baleful goddess and were incinerated by a blast of molten metal.

Climbing out of the crater, Douglas suddenly felt the rocks beneath him quake and shake from internal convolutions. It was a frightening experience: "Of all sensations in nature," he entered in his journal, "that produced by earthquakes or volcanic agency is most alarming. The strongest nerves are unstrung and the most courageous mind feels weakened and unhinged when exposed to either."

He spent still a third day at the crater's edge, then, suddenly remembering his expensive retinue which he was supporting in idleness, he made a quick trip up Mauna Loa (Big Mountain), was disappointed to find no sign of recent volcanic activity, and climbed back down again. He paid off his troupe and returned to Kilauea. "I must return to the volcano," he wrote. "I must return . . . if it is only to look–to look and admire."

As he looked by the hour, almost hypnotized, he began to read mystic revelations into the boiling lava, which seemed directly connected with the molten core of the earth, allowing "a worshiper at the portal of nature's temple," as he put it, "to pass within, and to be . . . a partaker of her mysteries." He came to believe that the spectacle before him, "which fills the mind with awe," was no "discordance of nature" but a means by which "that Being, in whom all truth, of whatever kind, finds its proper lasting place," was trying to communicate with human "ignorance." In time, if it pleased that Being, "all will be reconciled, and we shall see no longer as 'through a glass, darkly,' the infinite, the beauty, the harmony of nature." Even though the phrasing is derivative, the frame of mind is his own.

Time and again Douglas tries to capture in words the quality of the heaving, straining molten mass within its crater. "A vast sunken pit . . . with almost perpendicular sides" is one description of the enclosure; within the pit lies the "fiery lake, roaring and boiling in fearful majesty."

He keeps coming back to the image of a trapped force trying to break out of its confinement. The cauldron heaves, "throwing out lava in a thousand forms . . . like the breaking up of a large river with ice." And again: "The roaring and agitated state of the crater" heaved up and rolled around molten lava "in tremendous masses." And always he keeps us aware of the unearthly noises and the trembling of the hollow rock "faithless beneath [the] tread."

The imagery, the sense of entrapment, the struggle to break out, are so clearly delineated that we wonder if Douglas sensed a parallel to his own situation. Not long before, he had felt himself ready to overflow the world; now, flung back, turn or twist as he might, he had no place to go.

Possibly once again, at a different time and place, he felt a similar identification. The recognition had to be abrupt; he had no time to prepare for it. But on that morning to come on the Kohala highlands a sense of parallel fate may have drawn him irresistibly toward an animal trapped in another pit, a wild bull weighing half a ton, heaving and roaring and shaking the ground in frustration and fury.

Douglas returned to Honolulu in April and took up residence again with Richard Charlton, whom he described as "a most amiable and excellent man." It was an opinion not shared by American missionaries on Oahu, one of whom characterized the consul as a "beefy, red-faced Britisher, loud and aggressive . . . a bete noir to all decent or quiet people in Honolulu."

This derogatory opinion probably resulted from Charlton's refusal to suppress the "roaring hells" of the seaport. His attitude was that sailors on shore leave had a right to entertain themselves in their own way, a standard of tolerance which won little approval among tight-lipped New Englanders.

Douglas, finally convinced that time had run out on him, was on the lookout for a ship to take him back to England. He hoped for a westbound vessel that would carry him on around the world *via* the Indian Ocean and the Cape of Good Hope, thus enabling him to salvage something from the wreck of his plans to circle the globe by way of Siberia; but he was realistic enough to know that he couldn't be choosy and must take the first British craft putting into harbor.

Meanwhile, he worked at mounting and annotating the more than two thousand fern specimens he had collected on Hawaii and packing them for shipment to Professor Hooker. He also, and probably for the first time in his life, read for pleasure. It is doubtful if he had, since he first apprenticed as a gardener, put his nose inside a book—apart from the Bible—that had no bearing on his profession. A possible exception might have been the works of Edmund Burke, especially the celebrated essay, *On the Sublime and Beautiful.* It is reasonable to suppose that some more literate companion, perhaps John Lindley in the early Chiswick days, had stimulated Douglas to read the essay as an indispensable guide to an appreciation of nature. Certainly he seems to have taken to heart Burke's central precept that things which "tend to fill the mind with . . . delightful horror" are the "truest test of the sublime." With that clue, Douglas throughout his career had been able to see horror (or "awefulness" or "terror") in mountains, lakes, rivers, even trees; and most specifically, as we have seen, in volcanoes.

He had kept abreast of scientific journals as they reached him, often years late, and was a devoted reader of narratives of travel and exploration, especially works by his little coterie of heroes, Captains Vancouver, Franklin, and Beechey. But it is possible that until that spring in Honolulu, with time on his hands, his optical troubles cleared up and the Charlton ladies helpfully advising him, he had never read a novel, not even a novel by Sir Walter Scott, then rightly called "the most famous man alive, known all over the continents of Europe and America."

We can well believe that Scott's tales of honor and high purpose reached through to Douglas's starved sense of the romantic with shattering impact, and that he reacted much as a teen-ager when exposed to lofty sentiments for the first time, by shutting himself away for hours on end, lost in that glittering world of noble aims and achievements.

Spring blended into summer without visible change in the tropical landscape. There was still no ship for England, and Douglas, despite the companionship of Sir Walter Scott, was growing restless. The sails that finally appeared belonged to a bark from the United States, bringing in as passengers the Reverend John Diell from Hamilton College, New York, his wife, his child, and his enigmatic Negro manservant John.

As chaplain of the American Seamen's Friend Society, the Reverend Diell was to join in the crusade to clean up Honolulu, but since, at the time of his arrival, the waterfront was as prim as a New England village owing to the whaling crews being all at sea, he decided to utilize the lull by traveling around the islands. Possibly the trip had suggested itself to him as a result of meeting and conversing with Douglas about the volcanic origins of the archipelago. Diell, like many men of the cloth, was likely an amateur geologist. If so, he would have been fascinated by Douglas's description of the fire and brimstone activity in the crater of Kilauea, and must have considered himself most fortunate when Douglas consented to be his guest and guide on a visit to the volcano. Douglas, for his part, would feel almost as fortunate in getting away from the British consulate for a while so as not to add to his debts. He was bored by inaction and sick of waiting for a ship to show. Perhaps, neurotically, he was beginning to believe that His Majesty's sea captains were purposely avoiding Honolulu because they were aware of his status as a bankrupt pariah.

So it was that Douglas returned to the island of Hawaii to become entangled in an unanticipated sequence of events, beginning with his impulsive landing at Kohala Point, followed by the tramp across the highlands which had proved too much for John's feet and made his jettisoning necessary, and ending with the encounter with Edward Gurney, who volunteered to show Douglas the pits in which he captured cattle. Then there was the final scene when Douglas, turning moody and withdrawn, dismissed Gurney and proceeded on toward the pits alone.

What were his thoughts on that morning of July 12, 1834, as he began striding off the last miles of his life? We may assume that he got rid of Gurney because of irritation at himself for wasting time in chatter with an illiterate hunter and so, once alone, he no doubt made a special effort to lift his mind above the slaughter of cattle (he'd had enough of *that* in California, in all conscience). To seek what he called "the high capacities of our nature," it was his habit to give himself up to ennobling images, and now perhaps he set himself to contemplate the lives of heroes from the books he had been reading. Or, because Douglas was always of a practical turn of mind, he may have chosen to contemplate real heroes, of which his age had a bountiful supply, beginning with the sainted figure of Horatio, Lord Nelson.

Douglas was aware that much of Nelson's claim to immortality lay in his consummate timing of two factors, victory at Trafalgar and his own death at the very moment of it—a happy combination of events that suited perfectly the romantic mood of the times; so perfectly, in fact, that it set a pattern for dramatic climax which subsequent candidates for renown were expected to follow. And it wasn't always easy. Take the Duke of Wellington: He'd had the bad luck to survive Waterloo and live on to a grumpy, gout-ridden old age. Though he was always a popular figure, and even gained in veneration as his hair whitened, he was never quite forgiven by the British public for surviving his climactic achievement.

In truth, the nation as a whole seemed in love with death. There was a cachet about a well-timed expiration which appealed powerfully and perhaps nostalgically to a population settling down to a long night of industrial ignominy. It supplied a romantic aura that could enhance even the reputations of poets. To the man in the street, who cared nothing about flights of fancy, the names of Keats, Shelley, and Byron were well known because they belonged to men who died prematurely in distant lands, probably under adventurous circumstances. And for a man of action who wished to be remembered, romantic extinction was the almost indispensable termination for a career. In fact, death could sometimes obliterate from popular memory a shortage of actual achievement during life. Thus Mungo Park, an explorer who had hardly discovered anything, won undying fame by walking into darkest Africa and not walking out again.

The reliable Edmund Burke summed up precisely what the age expected of its heroes in an essay on General James Wolfe. One especially famous passage was often memorized by young officers in the armed forces as a prod and inspiration. Douglas probably didn't go that far, but beyond question he was familar with the text. For General Wolfe was, after all, a kind of patron saint of the Hudson's Bay Company since, by his victory over Montcalm at the Battle of Quebec in 1759, he eliminated the French from the beaver trade and made it once again safe for monopoly.

> "The death of Wolfe," wrote Burke, "was indeed grevious to his country, but to himself the most happy that can be imagined, and the most to be envied by all those who have a true relish for military glory. Unindebted to family or connections, unsupported by intrigue or faction, he had accomplished the whole business of his life when others were only beginning to appear; and at the age of thirty-five, without feeling the weakness of age, or the vicissitudes of fortune, having satisfied his honest ambition, having completed his character, having fulfilled the expectations of his country, he fell at the head of his conquering troops, and expired in the arms of victory."

The passage would have seemed of special significance to Douglas that July 12, 1834, because he had just passed his own thirty-fifth birthday.

There was also another kind of death facing Douglas as he walked across the highlands that he may have found even more distressing than the prospect of physical extinction. He had received (likely in Honolulu in 1832) a copy of Charles Lyell's *Principles of Geology* (first volume published 1830), probably sent to him by Professor Hooker. Lyell had revised and developed the concept of uniformitarianism, which held that the manifest changes that have occurred in the world throughout time came about by gradual processes—mountains crumbling through the agencies of weather and rivers, continents rising and subsiding through thrust and pull of volcanic forces. Lyell further maintained that nothing had happened during the endless millions of years of geologic history which could not be seen happening at the present time.

The hypothesis was deeply disturbing to orthodox geologists for several reasons: it drove a last nail into the coffin of the Genesis time table (from which Bishop Ussher of Armagh had calculated that Creation was completed at 8 P.M. Saturday, October 22, 4004 B.C.); it disposed of the "catastrophic" theory of change through cataclysmic floods or world-shattering earthquakes; finally it held that species, as evidenced by fossils in rocks, were in some way related to earlier, more primitive forms, and had not been created afresh by divine providence after each obliterating disaster. It was this last tenet that caused the greatest concern because of implications of transmutation ("evolution") in direct contradiction to the biblical mandate that fruits, flowers, whales, and serpents remain fixed "after their kind."

It is true that Lyell, always sensitive to public opinion, never suggested this development, but the inference was there and, long before Darwin pushed it to its logical conclusion, traditionalists were up in arms against it. One such was the Reverend Adam Sedgwick of Cambridge University who, in a speech before the Geological Society in 1831, assailed Lyell's concepts in uncompromising terms. He sneered at the doctrine of gradual change as assuming "that in the laboratory of nature, no elements have ever been brought together which we ourselves have not seen combined," then asked with increasing scorn, "And what is this but to limit the riches of the kingdom of nature by the poverty of our own knowledge, and to surrender ourselves to a mischievous, but not uncommon philosophical scepticism, which makes us deny the reality of what we have not seen and doubt the truth of what we do not perfectly comprehend?"

To David Douglas, reading these words in the "Proceedings" of the Geological Society, which reached him at Fort Vancouver, Sedgwick emerged as a champion of the faith, the wielder of a fiery sword against Lyell and other would-be destroyers of the old-time scientific religion. He attempted to memorize key passages of the address (garbling them more than once in letters) and tried to take comfort in the fact that, for the time being at least, heresy was being shouted down. But doubts must have remained. Lyell's argument, supported by telling facts and careful observations of nature, was too compelling to be dismissed. Even as Douglas ran from the logic, as unquestionably he did, he must have fired an indignant rearguard volley of questions.

What was happening in England to allow such impious ideas to be circulated? Was Sedgwick's the only voice raised in protest? Why were such stalwarts as Professor Hooker not only reading Lyell's book, apparently with approval, but sending it to disciples without apology? Inevitably—and this could have been shattering—other questions followed. Or was it he, David Douglas, who in his remoteness and isolation had fallen behind? Were new currents of temporal thought, divorced from religion, sweeping the scientific establishment? If so, what did it mean for his future? Was all his work, based on the assumption that God had clothed the world with plants in fulfillment of his covenant with man, now obsolete?

Douglas had never been "religious" in more than a formal sense during the years of his great achievements. But recently, especially since his arrival in Hawaii, he seemed to have traveled toward some mystic state. On the summit of Mauna Kea, as we have seen, he invoked a union with "a great and good, and wise and holy God," and the hours he spent at the brink of Kilauea's crater, viewing the work of "that Being, in whom all truth, of whatever kind, finds its proper lasting place," verged on religious ecstasy. We can only presume that as the fabric of his worldly ambitions fell in shreds he turned inward in an attempt to find some kind of spiritual compensation. He felt he could depend on the eternal values promised by the scriptures and venerated by men since the beginning of time. Yet here suddenly was Lyell, tearing down the veils, intent on revealing the mysteries, rewriting the Word, as it were, without opposition.

It must have seemed to Douglas that morning as he walked across the high shoulder of the mountain, a speck in an enormous landscape of rocks and sea and sky, that he was cut off from still another goal, perhaps the ultimate goal of one seeking security in a divine "lasting place." In his loneliness and isolation, drawn in upon himself, he may have felt like a man trapped in a corridor, hearing iron doors clang in succession, knowing each had sealed off an avenue of hope. Duty, loyalty, and heroism had failed him; now the very stability of God's world, to the glorious proof of which he had dedicated his working life, was crumbling under him.

In his despair, his mind working in circles ever more concentric, he may have plumbed the depths to grasp at a solution. The determination, once kindled, would burn bright, consuming doubts, religious prohibitions, and instincts of preservation in its flames. There was still a way out, how or when he might choose to take it. He was still master of his destiny, capable of demonstrating his contempt for a world he no longer understood by quitting it with honor.

Looking up, Douglas saw the cliffs, recognizing the rock formation Gurney had pointed out. Perhaps he already heard the bellow of the trapped bull, protesting its captivity and shaking the porous ground in its efforts to heave free. Douglas may have had a last look around, at the sun, at the sky, at the rounded rim of the ocean rolling over the horizon into infinity. He had at most twenty minutes of life remaining.

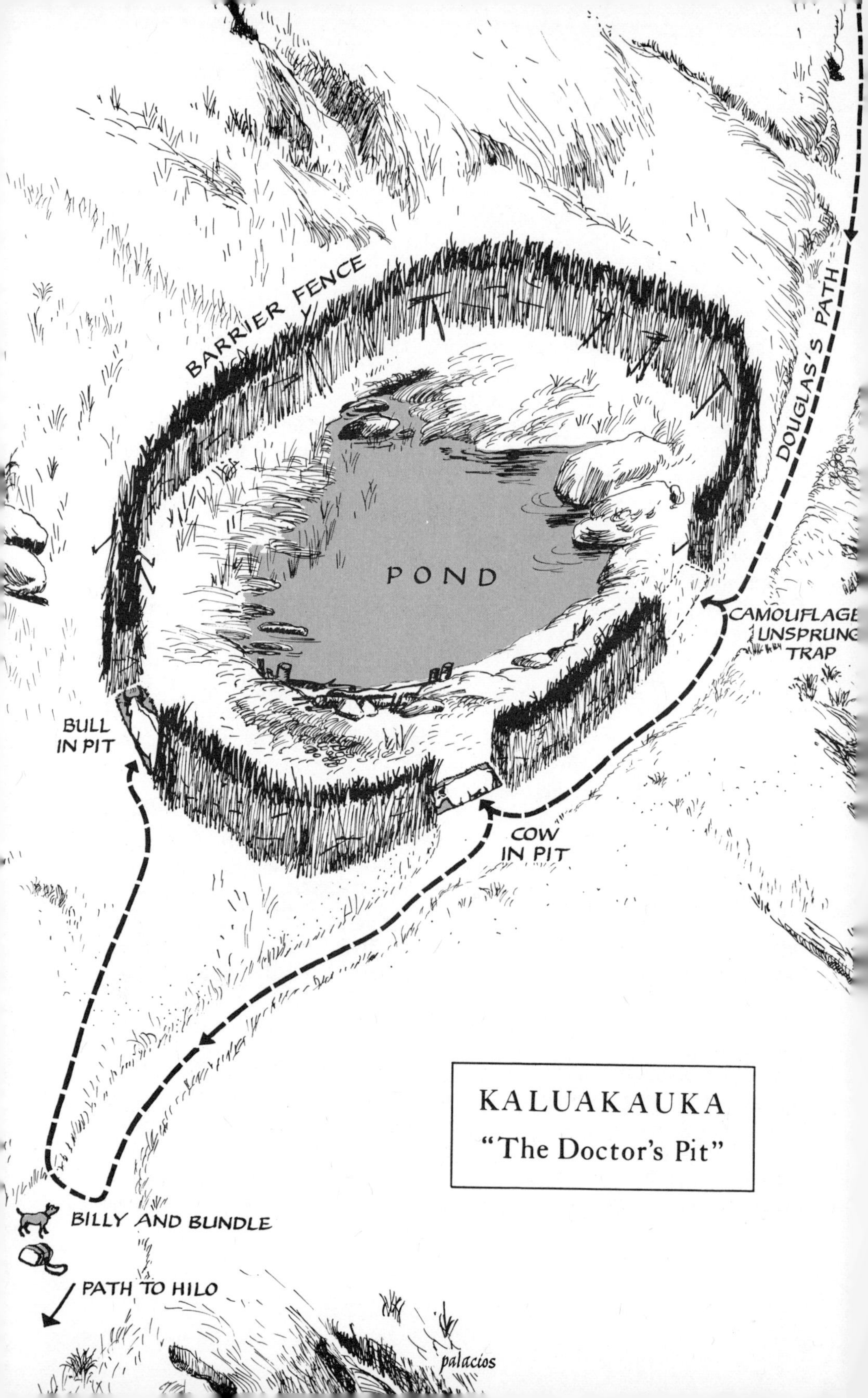
BARRIER FENCE
DOUGLAS'S PATH
POND
CAMOUFLAGE
UNSPRUNG
TRAP
BULL
IN PIT
COW
IN PIT
KALUAKAUKA
"The Doctor's Pit"
BILLY AND BUNDLE
PATH TO HILO
palacios

EPILOGUE

EDWARD GURNEY MUST STILL HAVE been incredulous as he hurried toward the pits. He had been so specific in his warnings; Douglas knew exactly what to expect; there must be some mistake. Then he saw his men chattering around the open trap and knew that somehow it had happened.

Below them the bull, weary but stirred to action again by the excitement around him, snorted and thrashed about, shaking the ground. Gurney could see a patch of clothing peeping out at the bottom of the pit, then feet, "the rest of the body being covered by dust and rubbish." He loaded his rifle, aimed behind the bull's horns, and fired. When the animal slumped, ropes were passed around the carcass and it was dragged from the pit.

The remains of David Douglas were hoisted up to Gurney who laid them on a cured hide he had brought for the purpose. The hat and the walking stick were dug out of the debris, but there was no sign of the bundle Douglas had carried on his shoulders. Gurney ordered a search, and after a few minutes the men heard a dog barking. Billy, tense, trembling but combative, was guarding his dead master's possessions about thirty yards down the path to Hilo.

The position of the dog and bundle in relation to the pits puzzled Gurney, so he set about tracing Douglas's movements. He was an experienced tracker and his work was made easier because of the heavy, nailed boots Douglas had been wearing. He had approached, as expected, from the north, stopped to examine the camouflaged, unsprung trap, then proceeded to the open pit holding the cow. After a pause there, Douglas left the scene to move on slowly down the path to Hilo, apparently listening, as he walked, to the bellowing bull.

Thirty yards down the path, Douglas had halted. He had slipped off his travel pack, leaving it in charge of Billy, and walked back to the bull pit, where the tracks terminated.

Gurney wrapped the body in the hide and had it carried to his lodge, where he made an inventory of the articles recovered. Then he accompanied the pallbearers down to the coast. Here he made a serious blunder. He had already been put to considerable trouble and expense (the pallbearers had to be paid and fed), so he might as well have continued to Hilo with the body, told his story, and been done with it. But apparently shrinking from contact with the missionaries, he entrusted Douglas's remains to a native boatman while "promising to come himself" within a matter of days with the dead man's dog and his personal effects.

The boatman put out to sea on the morning of July 14 and, propelled by the wind, which had changed direction, rapidly covered the thirty miles to Hilo. The same favorable wind had brought the Diells in from Molokai sooner than expected and they had already taken up residence with the Goodriches. The missionaries "were every moment awaiting [Douglas's] arrival" when, instead of the "*living* friend," a "mangled corpse" was delivered.

On recovering from their first "agony of feeling," Diell and Goodrich began preparing the body for burial. As they worked, they asked questions. How could Douglas, who had conquered mountains and glaciers throughout the western hemisphere, and most notably here in Hawaii, possibly fall by accident into a pit occupied by a raging bull? If the area was dangerous, why had Douglas been permitted to travel through it alone? Why hadn't Gurney accompanied him? For that matter, why hadn't John? And where was John now? The train of questions, once started, had no end.

They were appalled by the condition of the body. They counted "ten to twelve gashes on the head—a long one over the left eye, another, rather deep, just above the left temple, and a deep one behind the right ear." The ribs on the left side were broken, "the abdomen was also much bruised, and also the lower parts of the legs." The missionaries were inexperienced in acts of violence, but the circumstances of Douglas's death, combined with the multiple wounds on his head, suggested at least the possibility of murder.

Fortunately, or unfortunately, they believed they had an authority close by who could clear up the matter, a jack-of-all-trades called Hall whom they had engaged to dig Douglas's grave in Goodrich's churchyard. Among other things, Hall claimed to have been a cattle hunter. He examined the body carefully and then announced that he had never seen such lacerations inflicted by a bull's horns. In his opinion Douglas had been done to death by blows from an axe or some similar weapon and the body tossed into the pit to cover up the crime.

By now convinced that further investigation was necessary before consigning the body to the grave, Diell and Goodrich moved on two fronts; they dispatched Hall up the mountain to make inquiries while they, deeming "it due to the friends of Mr. D., and to the public, whom he had so zealously and so usefully served, that an examination be made of the body by medical men," eviscerated the corpse, filled it with salt, then placed it in a box filled with brine which they dispatched to Consul Charlton in Honolulu together with a detailed explanatory letter.

Up on the slopes of Mauna Kea, Hall found the pits, though heavy rains had wiped out footprints and other traces of activity. He interviewed Gurney, however, and learned that Douglas had mentioned leaving the missing black man somewhere along the trail. Stimulated by this clue, Hall retraced Douglas's steps across the highlands and uncovered Davis, the hunter with whom Douglas had spent his last night. It was then that Davis remembered, whether prompted by Hall or not, that Douglas had in his possession a large purse stuffed with money.

So Hall returned to Hilo with a motive for murder in the stuffed purse and also bringing with him an exhibit which he considered of the utmost importance, the head of a bull which he claimed came from the animal that had shared the pit with Douglas. He insisted that this gruesome object be also sent to Honolulu so the doctors could see that the lacerations on the body could not have been inflicted by horns. For some reason Hall seemed unable to visualize the cramped coffin of the pit where the bull was unable to lower its head to its victim.

While Hall was still on the mountain, Edward Gurney, alerted at last to his danger, came down to Hilo. He brought with him Douglas's personal effects and the little dog.* He told and retold his story to the missionaries, recounting the events of July 12 in such convincing detail that Diell and Goodrich finally confessed that their minds were "greatly relieved as to the probable way in which the fatal event was brought about."

But the relief proved only temporary. When Hall returned with his gory bull's head and his motive for murder, they became agitated all over

* Sent back to England, Billy found a new home with a clerk in the British Foreign Office.

again. Gurney's reputation as an escaped convict stood as a black mark against him, coupled with, of course, his unsanctified family arrangements, which were always distressing to missionaries. Still, when Hall proposed that they arrest Gurney on suspicion, Diell and Goodrich demurred, preferring to wait until a medical opinion was returned from Honolulu.

Gurney probably realized that, no matter what the verdict, he would never survive the shadow of guilt and his days on the mountain were numbered. In fact, within a few years he was forced off the islands as an undesirable. Legend has it that he was seen in California during gold rush days and then dropped into oblivion.

Owing to head winds, it took two weeks for the disemboweled body of David Douglas to reach Honolulu, arriving "in a most offensive state." Consul Charlton called in two civilian physicians to conduct the postmortem, and they were joined by two surgeons from H.M.S. *Challenger,* a British vessel which had arrived in port within days of Douglas's departure. This formidable battery of medical authorities delivered a unanimous opinion that all the injuries could have been inflicted by a bull's hoofs and that they saw no need to search farther for cause of death.

Charlton made arrangement for burial immediately after the postmortem and a final bizarre note was added to the circus atmosphere that had surrounded Douglas's death. In respect for the celebrated deceased, the interment was a starched affair, the service conducted by the chaplain of the *Challenger,* with the captain and his senior officers in full dress attendance. Only after the ceremony was it discovered that a mixup had occurred and that Douglas had been buried in the "native" section of the graveyard.

In the end it didn't matter. The brick paving Charlton ordered constructed above the plot soon disintegrated and the location was completely lost. A monument raised in 1855 was also eaten away by the climate. When the Horticultural Society finally got around to erecting a plaque, it was prudently mounted indoors in the vestibule of the church. Other memorials to Douglas's memory are to be found on the slope of Mauna Kea, at the site of his death, and in the churchyard of his birthplace, Scone. An inscription on the latter reads in part; "Endowed with an acute and vigorous mind, which he improved by diligent study, [he] uniformly exemplified in his conduct those Christian virtues which invested his character with a higher and more imperishable distinction than he justly acquired by his well-earned reputation for scientific knowledge."

Douglas would have been amazed to learn that his "Christian virtues," presumably meaning his personal qualities, could be extolled above his botanical achievements. For he believed that it was his inadequacy in these values, and his subsequent inability to lead men or control events, that defeated him.

It is easier for us today than it was for Douglas's contemporaries to understand his conflicts and ambiguities. We recognize that he was suffering from a neurosis we thought unique to our own times of judging himself by his failures rather than his successes. Few minds, then as now, can hold up under that strain.

Measured by his successes, we may think of David Douglas as his own generation thought of him, as a man of unique talents and restless energy to whom every lover of plants is indebted. Whatever his private anguish, he was able to find peace in nature, to journey with delight through a primeval landscape no longer available to us. As he once wrote; "It is a barren place that does not afford me a blade of grass, a bird or a rock . . . from which I derive inexpressible delight." We who seem at times to have reached that barren place might well envy him.

SELECTED BIBLIOGRAPHY

Allan, Mea. *The Hookers of Kew, 1785–1911.* London: Michael Joseph, 1967.

Allen, B. Sprague. *Tides in English Taste: A Background for the Study of Literature.* Vol. 2. Cambridge: Harvard University Press, 1937.

Anderson, A. W. *How We Got Our Flowers.* New York: Dover Publications, 1966.

Anderson, Edgar. *Plants, Man and Life.* Boston: Little, Brown, 1952.

Arnold Arboretum. "Notes on Making a Herbarium." *Arnold Arboretum Publications* 28, nos. 8 and 9. Jamaica Plain, Mass.: Arnold Arboretum Publications, 1968.

Atherton, Gertrude. "Concha Argüello, Sister Dominica." *The Spinners' Book of Fiction.* San Francisco: Paul Elder, 1907.

———. *California: An Intimate History.* New York: Harper & Bros., 1914.

Bakeless, John E. *The Eyes of Discovery: The Pageant of North America as Seen by the First Explorers.* Philadelphia: J. B. Lippincott Co., 1950.

———. *Lewis and Clark, Partners in Discovery.* New York: Wm. Morrow & Co., 1947.

Ballantyne, Robert Michael. *Hudson Bay; or, Everyday Life in the Wilds of North America.* London: T. Nelson & Sons, 1896.

Bancroft, Hubert Howe. *California Pastoral, 1769–1848.* San Francisco: The History Co., 1888.

Bancroft, Hubert Howe. *History British Columbia, 1792–1887*. San Francisco: The History Co., 1887.

———. *History of California*. Vols. 1, 2, 3, 4, and 5. San Francisco: The History Co., 1884–90.

———. *History of the Northwest Coast*. Vols. 1 and 2. San Francisco: A. L. Bancroft & Co., 1884.

———. *History of Oregon*. Vols. 1 and 2. San Francisco: The History Co., 1886–88.

———. *History of Washington, Idaho and Montana*. San Francisco: The History Co., 1890.

———. *History of the North Mexican States and Texas*. Vols. 1 and 2. San Francisco: A. L. Bancroft & Co., 1884–89.

Bandini, José. *A Description of California in 1828*. Berkeley: Friends of the Bancroft Library, 1951.

Barnston, George. "Abridged Sketch of the Life of Mr. David Douglas, Botanist, with a few Details of His Travels and Discoveries." *The Canadian Naturalist and Geologist* 5 (1860): pp. 120–32, 200–208, 267–78, 329–49.

Bates, Marston. *The Forest and the Sea: A Look at the Economy of Nature and the Ecology of Man*. New York: Random House, 1960.

Beechey, F. W. *Narrative of a Voyage to the Pacific . . . in the Years 1825, 1826, 1827, 1828*. Vols. 1 and 2. London: Henry Colburn & Richard Bentley, 1831.

———. *An Account of a Visit to California, 1826–27*. San Francisco: The Grabhorn Press, 1944.

Bishop, Sereno Edwards. *Reminiscences of Old Hawaii*. Honolulu: Hawaiian Gazette Co., 1916.

Blair, Thos. Arthur. *Weather Elements: A Text in Elementary Meteorology*. New York: Prentice-Hall, 1937.

Bourinot, John, ed. "Diary of Nicholas Garry, Deputy-Governor of the Hudson's Bay Company from 1822–1835." *Proceedings and Transactions of the Royal Society of Canada*, 2nd series, vol. 6, sec. 2 (1900): 72–204.

Bovill, Edward William. *English Country Life, 1780–1830*. London: Oxford University Press, 1962.

Bryan, L. W. "Kaluakauka." *Paradise of the Pacific* (Honolulu), December 1934, pp. 28–30.

Bryant, Arthur. *The Age of Elegance, 1812–1822*. London: Wm. Collins Sons & Co., 1950.

Buchan, John. *Sir Walter Scott*. 7th ed. London: Cassell & Co., 1946.

Burke, Edmund. *On the Sublime and Beautiful*. The Harvard Classics, vol. 24. New York: P. F. Collier & Son, 1909–10.

———. *The Annual Register, or a View of the History, Politics and Literature for the Year 1759* (p. 41 for an account of Wolfe's death). 8th ed. London: J. Dodsley, 1792.

Burt, Alfred LeRoy. *A Short History of Canada for Americans*. Minneapolis: The University of Minneapolis Press, 1944.

Burton, Elizabeth. *The Pageant of Georgian England*. New York: Chas. Scribner & Sons, 1967.

Byers, Horace Robert. *General Meteorology*. 3rd ed. New York: McGraw-Hill, 1959.

Cameron, Hector Charles. *Sir Joseph Banks, the Autocrat of the Philosophers*. London: Batchworth Press, 1952.

Carey, Charles H. *A General History of Oregon Prior to 1861*. Vols. 1 and 2. Portland: Metropolitan Press, 1935–36.

Carlyle, Thomas. *Sir Walter Scott*. The Harvard Classics, vol. 25. New York: P. F. Collier & Son, 1909–10.

———. *On Heroes, Hero-Worship and the Heroic in History*. The World's Classics, vol. 62. London: Oxford University Press, 1963.

Chancellor, E. Beresford. *Life in Regency and Early Victorian Times: An Account of the Days of Brummell and D'Orsay 1800–1850*. London: B. T. Batsford, 1926.

Claiborne, Robert. *Climate, Man and History*. New York: W. W. Norton & Co., 1970.

Coats, Alice M. *The Plant Hunters: Being a History of the Horticultural Pioneers, Their Quests and Their Discoveries from the Renaissance to the Twentieth Century*. New York: McGraw-Hill, 1970.

———. *Garden Shrubs and Their Histories*. New York: E. P. Dutton & Co., 1965.

Coues, Elliott, ed. *History of the Expedition Under the Command of Lewis and Clark to the Pacific Ocean . . . During the Years 1804–5–6*. Vol. 2. New York: F. P. Harper, 1893.

Coulter, John M. *Fundamentals of Plant Breeding*. New York: D. Appleton & Co., 1925.

Coulter, Thomas. *Notes on Upper California: A Journey from Monterey to the Colorado River in 1832*. Early California Travel Series, no. 1. Los Angeles: G. Dawson, 1951.

Cox, Ross. *The Columbia River; or, Scenes and Adventures During a Residence of Six Years on the Western Side of the Rocky Mountains . . . Together with a Journey Across the American Continent*. The American Exploration and Travel Series, no. 24. Norman: University of Oklahoma Press, 1957.

Dakin, Susanna Bryant. *The Lives of William Hartnell*. Palo Alto: Stanford University Press, 1949.

Dana, R. H. *Two Years Before the Mast.* The Harvard Classics, vol. 23. New York: P. F. Collier & Son, 1909–10.

Davis, William Heath. *Sixty Years in California: A History of Events and Life . . . Under the Mexican Regime . . . and After the Admission of the State into the Union.* San Francisco: A. J. Leary, 1889.

DeRos, John F. F. *Travels in the United States and Canada in 1826.* 3rd ed. London: W. H. Ainsworth, 1827.

Dillon, Richard. *Meriwether Lewis: A Biography.* New York: Coward-McCann, 1965.

Douglas, David. *Journal Kept by David Douglas During his Travels in North America, 1823–1827, Together with . . . Appendices Containing a List of the Plants Introduced by Douglas and an Account of His Death in 1834.* Published under the direction of the Royal Horticultural Society. London: William Wesley & Son, 1914. Reprint in fascimile. New York: Antiquarian Press, 1959.

Douglass, A. E. *Climatic Cycles and Tree Growth.* Washington, D.C.: Carnegie Institution of Washington, 1919.

Drucker, Philip. *Indians of the Northwest Coast.* American Museum Science Books, no. B3. New York: Natural History Press, 1963.

Drummond, Thomas. "Sketch of a Journey to the Rocky Mountains and to the Columbia River in North America." In *Botanical Miscellany*, Vol. 1, edited by Sir William J. Hooker, pp. 178–219. London: John Murray, 1830–33.

Duhaut-Cilly, Auguste. "Duhaut-Cilly's Account of California in the Years 1827–28." *California Historical Society Quarterly* 8 (1929): 130–66, 214–50, 306–36.

Eifert, Virginia S. *Tall Trees and Far Horizons.* New York: Dodd, Mead & Co., 1965.

Eiseley, Loren. *Darwin's Century.* New York: Doubleday & Co., 1958.

Eldredge, Z. S., ed. *History of California.* Vol. 2. New York: The Century History Co., 1915.

Elliott, T. C., ed. "Journal of John Work, 1824." *Washington Historical Quarterly* 3 (1912): 206–7.

Engelhardt, Zephyrin. *The Missions and Missionaries of California.* San Francisco: The James H. Barry Co., 1916.

Ermatinger, Edward. "Edward Ermatinger's York Factory Express Journal; Being a Record of Journeys Made Between Fort Vancouver and Hudson Bay in the Years 1827–1828." *Proceedings and Transactions of the Royal Society of Canada*, 3rd series, vol. 6, sec. 2, (1912): 70–132.

Farb, Peter. *Man's Rise to Civilization, as Shown by the Indians of North America from Primitive Times to the Coming of the Industrial State.* New York: E. P. Dutton & Co., 1963.

Felton, Ernest L. *California's Many Climates.* Palo Alto: Pacific Book, 1965.

Fletcher, H. R. *The Story of the Royal Horticultural Society 1804–1968.* London: Oxford University Press, 1969.

Franchère, Gabriel. *Adventure at Astoria, 1810–1814.* Montreal, 1820 (in French). Reprint. Norman: University of Oklahoma Press, 1967.

Franklin, John. *Narrative of a Journey to the Shore of the Polar Sea, in the Years 1819, 1820, 1821, 1822.* Vols. 1 and 2. London: John Murray, 1823.

Gibbs, George. *A Dictionary of the Chinook Jargon; or, Trade Language of Oregon.* New York: Cramoisy Press, 1863.

Gillispie, Charles C. *Genesis and Geology.* Cambridge: Harvard University Press, 1951.

Gleason, H. A. and Cronquist, A. *The Natural Geography of Plants.* New York: Columbia University Press, 1964.

Goldie, John. *Diary of a Journey Through Upper Canada and Some of the New England States in 1819.* Toronto: (no publisher listed) 1897.

Good, Ronald. *The Geography of the Flowering Plants.* 3rd ed. London: Longmans, 1964.

Gourlie, Norah. *The Prince of Botanists, Carl Linnaeus.* London: H. F. and G. Witherby, 1953.

Graustein, Jeannette E. *Thomas Nuttall, Naturalist.* Cambridge: Harvard University Press, 1967.

Hadfield, Miles. *The Pioneer Plant Collectors.* London: Routledge and Kegan Paul, 1955.

Harvey, Athelstan George. *Douglas of the Fir.* Cambridge: Harvard University Press, 1947.

Hawaiian Mission Children's Society, pub. *Portraits of American Protestant Missionaries to Hawaii.* Honolulu: The Hawaiian Gazette Co., 1901.

Hawks, Ellison. *Pioneers of Plant Study.* London: The Sheldon Press, 1928.

Hawthorne, Hildegarde. *California Missions.* New York: Appleton-Century, 1942.

Holbrook, Stewart H. *The Columbia.* Rivers of America Series. New York: Rinehart & Co., 1956.

Hooker, William J. "A Brief Memoir of the Life of Mr. David Douglas, with Extracts from His Letters." *Companion to the Botanical Magazine* (London) 2 (1836): 79–182. Reprint. *Oregon Historical Quarterly* 5 (1904): 223–71, 325–69; 6 (1905): 76–97, 206–27, 288–309, 417–49.

Howell, John Thomas. "A Collection of Douglas' Western American Plants."

Leaflets of Western Botany (San Francisco) 2 (1937–40): 59–62, 74–77, 94–96, 116–19, 139–44, 170–74, 189–92.

Hutchinson, W. H. *California—Two Centuries of Man, Land and Growth in the Golden State*. Palo Alto: American West Publishing Co., 1969.

Irving, Washington. *Astoria*. New York, 1836. Reprint. New York: The Century Co., 1911.

Jepson, William Linn. "David Douglas in California." *Madroño, Journal of California Botanical Society*, 2, no. 12 (1933): 97–100.

Johnson, Robert C. *John McLoughlin: Patriarch of the Northwest*. Portland, Ore.: Metropolitan Press, 1935.

Kreig, Margaret B. *Green Medicine*. New York: Rand McNally & Co., 1964.

Lamb, G. F. *Franklin—Happy Voyager*. London: Ernest Benn, 1956.

Lamb, W. Kaye, ed. *Letters and Journals of Simon Fraser, 1806–08*. Toronto: Macmillan, 1960.

Lane, Ferdinand C. *Earth's Grandest Rivers*. New York: Doubleday & Co., 1949.

Langsdorff, George H. von. *Langsdorff's Narrative of the Rezánov Voyage to Nueva California in 1806*. San Francisco: the private press of T. C. Russell, 1927.

Lee, W. Storrs. *The Islands*. New York: Rinehart and Winston, 1966.

Leopold, Aldo. *A Sand County Almanac*. New York: Oxford University Press, 1949.

Lewis, William S., and Murakami, Naojiro, eds. *Ranald MacDonald: The Narrative of His Early Life on the Columbia Under the Hudson's Bay Company's Regime*. Spokane: Inland-American Printing Co., 1923.

Lindsay, T. S. *Plant Names*. London: The Sheldon Press, 1923.

Lorraine, M. J. *The Columbia Unveiled*. Los Angeles: Times-Mirror Press, 1924.

Lyell, Charles. *Principles of Geology; or, The Modern Changes of the Earth and its Inhabitants, Considered as Illustrative of Geology*. Vols. 1, 2, and 3. Reprinted from the 6th English edition. Boston: Hilliard, Gray and Co., 1842.

Lyman, Chester S. *Around the Horn to the Sandwich Islands and California, 1845–1850*. New Haven: Yale University Press, 1924.

Lyman, George D. "The Scalpel Under Three Flags." *California Historical Society Quarterly* 4 (1925): 142–206.

Lyman, Henry M. *Hawaiian Yesterdays*. Chicago: A. C. McClurg, 1906.

McArthur, Lewis A. *Oregon Geographic Names*. 3rd ed. Portland: Binsford & Mort, 1952.

Macartney, C. E., and Dorrance, G. *The Bonapartes in America*. Philadelphia: Dorrance & Co., 1939.

MacDonald, D. F. *Age of Transition*. New York: St. Martin's Press, 1960.

McLeod, Alexander Roderick. *The Hudson's Bay Company's First Fur Brigade to the Sacramento Valley: Alexander McLeod's 1829 Hunt*. Fair Oaks, Calif.: Sacramento Book Collectors Club, 1968.

Maloney, Alice B. "John Work of the Hudson's Bay Company, Leader of the California Brigade of 1832–33." *California Historical Society Quarterly* 22 (1943): 97–109.

Martin, Horace T. *Castorologia; or, The History and Traditions of the Canadian Beaver*. Montreal: Wm. Drysdale & Co., 1892.

Meany, Edmond S. *Origin of Washington Geographic Names*. Seattle: University of Washington Press, 1923.

Menzies, Archibald. "Menzies' California Journal." *California Historical Society Quarterly* 2 (1924): 265–340.

Merk, Frederick, ed. *Fur Trade and Empire; George Simpson's Journal, 1824–25*. Cambridge: Harvard University Press, 1968.

Montgomery, Richard G. *The White-Headed Eagle*. New York: The Macmillan Co., 1934.

Moorehead, Alan. *Darwin and the Beagle*. New York: Harper & Row, 1969.

Morgan, Dale L. *Jedediah Smith*. New York: The Bobbs-Merrill Co., 1953.

Morgan, Lewis H. *The American Beaver and His Works*. Philadelphia: Lippincott, 1868.

Munz, Philip A., and Keck, David D. *A California Flora*. Berkeley: University of California Press, 1959.

Murray, J. M. "David Douglas." *Forestry* (London) 5 (1931): 154–58.

Newcombe, Charles F., ed. *Menzies' Journal of Vancouver's Voyage, April to October, 1792*. Victoria, B. C.: William H. Cullin, 1923.

Oliver, F. W., ed. *Makers of British Botany*. Cambridge, England: University Press, 1913.

Platt, Rutherford. *The Woods of Time*. New York: Dodd, Mead & Co., 1957.

Putnam, Ruth. "California: The Name." *University of California Publications in History*, 4, no. 4 (1917), pp. 293–365.

Reed, Henry E. "William Johnson." *Oregon Historical Quarterly* (1933): 314–23.

Rich, E. E., ed. *Black's Rocky Mountain Journal, 1824*. The Hudson's Bay Record Society, vol. 18. London: The Hudson's Bay Record Society, 1955.

Roberts, George B. "Recollections of the Hudson's Bay Company." Unpublished Manuscript. Berkeley: Bancroft Library, University of California.

Robbins, Wilfred W. "Alien Plants Growing Without Cultivation in California." *University of California Agricultural Experimental Station,* Bulletin 637, July 1940.

Robbins, Wilfred W., and Ramaley, Francis. *Plants Useful to Man.* Philadelphia: P. Blakiston's Son & Co., 1937.

Robinson, Alfred. *Life in California.* New York: Wiley & Putnam, 1846. Reprint. San Francisco: the private press of T. C. Russell, 1925.

Rogers, Cameron. *Trodden Glory.* Santa Barbara: Wallace Hebberd, 1949.

Rojas, A. R. *The Vaquero.* Santa Barbara: McNally & Loftin, 1964.

Ross, Alexander. *Adventures of the First Settlers on the Oregon or Columbia River.* London, 1849. Reprint. Chicago: The Lakeside Press, 1923.

———. *The Fur Hunters of the Far West.* London, 1853. Reprint. Norman: University of Oklahoma Press, 1956.

Sandoz, Mari. *The Beaver Men.* New York: Hastings House, 1964.

Saunders, Charles Francis. *Western Wild Flowers and Their Stories.* New York: Doubleday, Doran & Co., 1933.

Scouler, John. "Dr. John Scouler's Journal of a Voyage to N. W. America." *Oregon Historical Quarterly* 6 (1905): 54–75, 159–205, 276–87.

Sheppe, Walter, ed. *First Man West: Alexander Mackenzie's Journal of his Voyage to the Pacific Coast of Canada in 1793.* Berkeley: University of California Press, 1962.

Smith, Edward. *The Life of Sir Joseph Banks.* New York: John Lane, 1911.

Spier, Leslie. *Tribal Distribution in Washington.* General Series in Anthropology, no. 3. Menasha, Wis.: George Banta Publishing Co., 1936.

Stewart, George R. "The Source of the Name 'Oregon.'" *American Speech* 19 (1944): 115–17.

Stroud, Dorothy. *Capability Brown.* London: Country Life Ltd., 1950.

———. *Humphrey Repton.* London: Country Life Ltd., 1962.

Sunset eds. *The California Missions: A Pictorial History.* Menlo Park: Lane Book Co., 1968.

Taylor, Norman. *Encyclopedia of Gardening, Horticulture and Landscape Design.* 3rd ed. Cambridge, Mass. The Riverside Press, 1956.

Thwaites, Reuben G., ed. *Nuttall's Travels into the Arkansa Territory, 1819.* Early Western Travels, 1748–1846, vol. 13. Cleveland: The Arthur H. Clark Co., 1905.

———, ed. *Original Journals of the Lewis and Clark Expedition 1804—6.* Vol. 3. New York: Dodd, Mead & Co., 1905.

Tyrrell, J. B., ed. *David Thompson's Narrative of His Explorations in Western America, 1784–1812.* Publications of The Champlain Society, vol. 12. Toronto: The Champlain Society, 1916.

Vancouver, George. *A Voyage of Discovery to the North Pacific Ocean and Round the World in the Years 1790, 1791, 1792, 1793, 1794 and 1795.* 3 vols. London: G. G. and J. Robinson, 1798.

Webber, Ronald. *The Early Horticulturalists.* Newton Abbot, Devon: David & Charles, 1968.

Wilbur, Marguerite Eyer, ed. *Vancouver in California, 1792–1794.* Los Angeles: G. Dawson, 1953.

Wilkes, Charles, U. S. N. *Narrative of the U. S. Exploring Expedition During the Years 1838, 1839, 1840, 1841, 1842.* 5 vols. Philadelphia: C. Sherman, 1844–74.

INDEX

Illustrations are indicated in italics

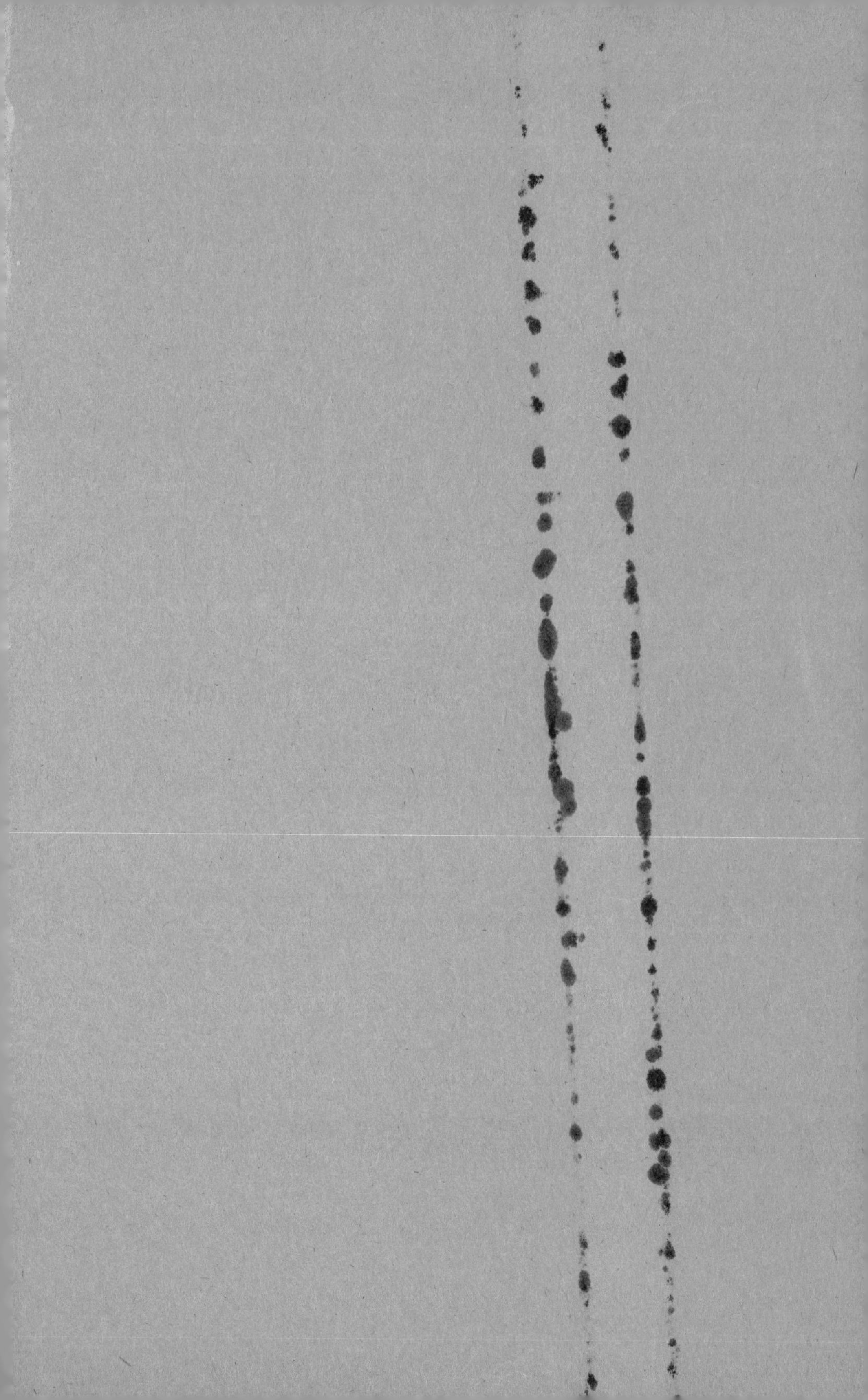

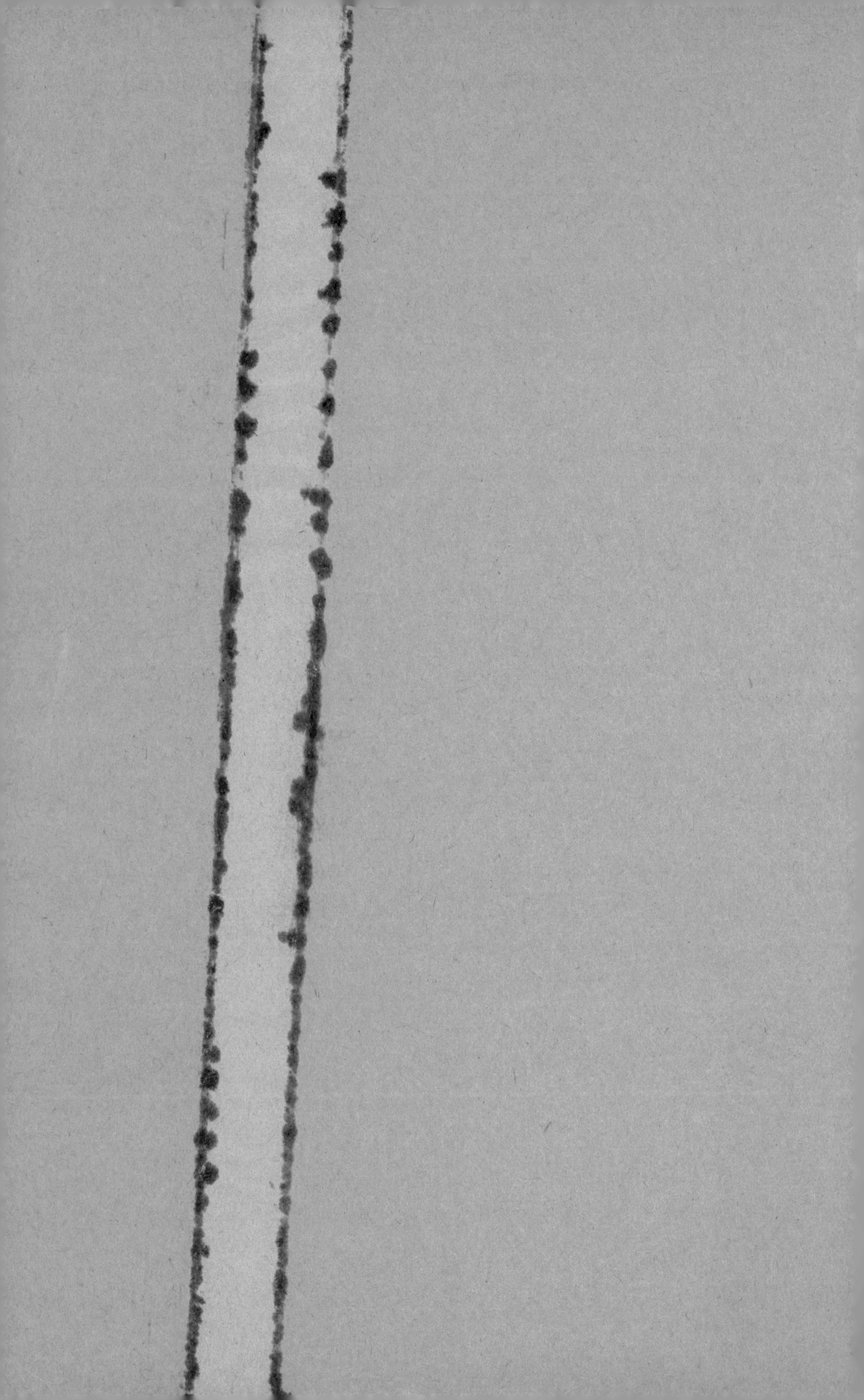